Parviz Khoshvaght

God Never Dies

Parviz Khoshvaght

God Never Dies

Physics and metaphysics everything theory

Blessed Hope Publishing

Imprint

Any brand names and product names mentioned in this book are subject to trademark, brand or patent protection and are trademarks or registered trademarks of their respective holders. The use of brand names, product names, common names, trade names, product descriptions etc. even without a particular marking in this work is in no way to be construed to mean that such names may be regarded as unrestricted in respect of trademark and brand protection legislation and could thus be used by anyone.

Cover image: www.ingimage.com

Publisher:
Blessed Hope Publishing
is a trademark of
Dodo Books Indian Ocean Ltd. and OmniScriptum S.R.L publishing group

120 High Road, East Finchley, London, N2 9ED, United Kingdom
Str. Armeneasca 28/1, office 1, Chisinau MD-2012, Republic of Moldova, Europe
Managing Directors: Ieva Konstantinova, Victoria Ursu
info@omniscriptum.com

Printed at: see last page
ISBN: 978-620-9-17354-7

In the name of the creator of the universe

GOD NEVER DIES

Physics and metaphysics everything theory

Author Name
Parviz khoshvaght

Translator

Nazanin Ehsani Tabatabaee

Where did it begin?

About sixty years ago, during my childhood, I became deeply motivated to understand the secrets of existence and the nature of God. This obsession consumed me; for many years, day and night, I pondered the countless "whys" that kept multiplying. After completing my primary education, I entered high school, where I immersed myself in the works of Greek, Eastern, and Western philosophers, studying their writings with relentless dedication. I also delved into physics books as I continued my quest for understanding. Ultimately, I realized that throughout history, countless individuals like myself have sought answers to the mysteries of existence, yet have passed on without finding them. However, my ambitions were much higher.

After years of research, I concluded that the only way to solve the mystery of existence is to understand the nature of time. No one in history has ever succeeded in doing this. After many years of effort, I finally managed to enter the realm of time, but this was just the beginning. I still had thousands of questions to answer. Although I grew tired of reading and studying, I did not lose hope. Eventually, I turned my focus to physics and quantum physics, exploring the universe itself. It was a great blessing from the intelligent network of the universe that nearly all the closed doors were opened to me, leading to the writing of this book that I now offer to you, humanity.

After years of research, I am sharing my insights with humanity in this book. I hope you will utilize them wisely. The content spans philosophical, physical, and scientific dimensions, which may be challenging to comprehend. For every hour you read, dedicate additional time to reflection. Contextualize these ideas within the universe for a deeper understanding and pursue further studies. I have learned that consciousness unfolds gradually through good deeds rather than being revealed all at once. Some may struggle to grasp the book's concepts, but that is not my fault. Seek awareness, and you will find that proper understanding comes with spiritual consciousness.

Humans differ significantly in their levels of spiritual awareness, and good deeds, not evidence or science, foster understanding. While I see the younger generation as realistic, many still lack proper awareness. The world is progressing towards enlightenment. I urge you to pursue reason and wisdom, especially logic—the path I have followed. Always maintain a critical mindset, questioning everything with a focus on reason, wisdom, and logic.

When offering criticism, seek to identify the right and logical solutions. Avoid being a hostile critic. If your criticism closes off specific paths, be sure to open the right ones as well. If you are unsure about something, dare to admit, "I don't know." By doing this, you will avoid fooling yourself. Awareness is a source of peace, and peace brings comfort. Pursue awareness, as it is an ongoing journey. Remember, today's actions shape tomorrow's destiny, so focus on doing good.

The fate of humanity lies in our own hands; God does not determine it. God serves as a guide, but the responsibility rests with us. Writing this book was a duty I felt compelled to fulfill, and I owe no favors to anyone for having done so. I understand that you, too, have questions—just as I once did—questions that may never be fully answered by others or even by yourself.

I hope that by reading this book, you will find the answers you seek and that the world will move toward enlightenment from this point forward. This process began on Earth a few years ago and is now accelerating. This book is critical and necessary during this crucial time in human history.

During my research, I continually sought to understand the reasons behind the system of existence, from day to night. This quest led me to explore the philosophical dimension, drawing me, willingly or unwillingly, towards quantum physics, the science that studies the structure of the universe. Consequently, I found it necessary to follow this path in conjunction with philosophy.

I utilized the theories of various scientists and physicists, with a particular focus on Einstein's theory of relativity. Unfortunately, I found that it did not provide a visualization or framework that resonated with my understanding of existence. As a result, I dedicated years to studying the system of existence and contemplating the relationships involved in the universe's emergence. Fortunately, this was the most illuminating path available to me during my research. I resurrected the science of physics, which had seemed to be submerged beneath the surface, and brought it to light. My first significant achievement was discovering the nature of time and formulating a mathematical equation for it—a concept that has perplexed humanity since the dawn of civilization. This research continued to unveil the twenty laws governing the structure of the universe, which I have outlined in this book.

Unfortunately, many theories from twentieth-century scientists lack scientific merit and do not fit within the framework of existence, effectively creating a barrier that has confined physicists. Not only do they resist criticism, but they also hesitate to step outside of this fence, a situation that has contributed to the current deadlock in physics.

It is both surprising and perplexing that the scientific community, especially those who deal with established science, remains closed off to criticism and

reluctant to adjust their course. The theories of the past deserve scrutiny and revision. I have often observed physicists exhibiting arrogance and a sense of superiority. They often develop a strong bias against past physicists, believing their theories to be divine revelations bestowed by a higher power.
Such biases and arrogance strike me as foolishness. Philosophically, pride and prejudice act as barriers that are difficult, if not impossible, to overcome, restricting human consciousness. As seekers of knowledge and understanding, we must remain open to insights from any source, whether it is an ordinary individual or a renowned scientist. As intelligent beings, we bear the responsibility to pursue the best pathways to understanding.
As a philosophical researcher and theorist, I assert that those who seek a meaningful way of life must welcome criticism and also practice critical thinking. They should not accept anything unquestioningly without careful consideration and rational, scientific reasoning.
I have conducted thorough investigations into metaphysics, which aligns with scientific principles. Our existence is grounded in the knowledge available to us. If we lack understanding about existence and metaphysics, it signals a gap in our knowledge, not a flaw in the system of existence itself. The primary imperfection lies within us. I stress that when humanity lacks knowledge in certain areas, it should not dismiss those areas entirely. Instead, we must boldly acknowledge that our understanding has not yet reached that point.
Wishing you victory, success, and knowledge.

Parviz Khoshvagt
Gohardasht, Iran

CONTENTS

PART 1: THE CHARIOT OF TIME, ITS ESSENCE, AND MATHEMATICAL FORMULA

It is no exaggeration to say that every thinker and scientist in the world contemplates the concept of time. Time, a complex force that has existed since the universe's inception, has shaped the world into its current magnificence. Scientists have observed approximately 95 billion light-years of space and have yet to reach the edge of the universe, which contains trillions of planets, stars, galaxies, and even more universes. Time has shaped the world's greatest wonders, encompassing all kinds of creatures, including animals, plants, and humans. Humans possess reason, intelligence, and the desire to explore and investigate. The first sensation a person experiences at birth is the passage of time, and their final awareness is also of time, which relentlessly pursues them. What is this time, and what power does it hold that it can eradicate all existence without the slightest hint of delay, rest, or negligence in its workings?

Time acts like an executioner wielding a sword that ruthlessly cuts apart and destroys everything—mountains, seas, oceans, planets, and stars. It brings all humans into the world, subjecting them to countless hardships: the old and weary, the young and vibrant, the ugly and beautiful, the strong and weak, the rulers and the submissive, as well as those condemned. Ultimately, it leads them all to death. Time is a burning fire that consumes all existence, indifferent even to the ashes it leaves behind, dragging everything toward destruction. It is an unbearable, roaring storm that leaves nothing untouched in the realm of existence, surrounding and drowning everyone in death and ruin. Can we call "Time" God? Time is not God, but it is an expression of God's power.

As long as you are alive, you eat, breathe, walk, engage in activities, rest, and sleep. Then, the cycle begins again. You live your life for weeks, months, and years until you eventually die. When you surrender your life to the Creator, time may seem to stand still for you, but in reality, it continues to pass. You either exist or you do not.

Now that you are alive, you experience feelings and often check your watch throughout the day, occasionally glancing at the calendar to keep track of the passing days, months, and years. You have plans for both your work and life, and you continue to meet with friends, family, and relatives. You may also find yourself reflecting on the past during the day. It's essential to remember that you are constantly surrounded by the flow of time, often without even being aware of it. The truth is, there is no escaping the influence of time. It envelops you, much like a fish swimming in the depths of the ocean.

What is time ?

This question has intrigued humanity since the beginning of our existence on Earth. Yet, a definitive answer must be found. The greatest philosophers and scientists have pondered and searched for answers to it for years. Yet, they have ultimately found no resolution before succumbing to the passage of time . In my opinion, humans possess seven senses. Alongside the traditional five senses, there is an additional sense I call the "inner sense," which allows us to experience hunger, thirst, fatigue, agility, and pain. However, there is yet another sense that serves as the foundation for all the others: reason. This cognitive ability has been ingrained in human existence, enabling us to understand and analyze the other senses. It helps us differentiate between good and evil, ugly and beautiful, truth and falsehood, justice and injustice, among many different concepts .Without this sense of reason, what would humanity be? Would we even have an actual existence? Thus, I argue that humans have seven senses.

The six senses function as the body's news agencies, delivering information to the brain, which stores it in its memory warehouse as long as the intelligence system is awake and conscious. The human mind then analyzes this information. This system also perceives additional stimuli and records them for timely retrieval. When you wake up in the morning, have breakfast, and leave the house at 8:00, you arrive at work with a plan or schedule that you created the night before. After finishing your workday, you return home, reflecting on the tasks completed and the events that occurred. During this reflection, you become aware of the passage of time. Have you ever considered what you review in your mind?

You have recorded the effects of your work, the events that transpired, and other incidents in your brain, allowing you to feel the passage of time. You revisit these memories while resting or whenever you need to. It's essential to note that during sleep or unconsciousness, you cannot perceive the passage of time, as your senses are inactive. Have you thought about this? In all the events and actions, whether performed by you or others, forces and energies are exchanged to bring these happenings into existence. You may not have considered this before, but it is true. Physics supports this notion: without a force, no work or movement can occur. Similarly, no movement or displacement can occur without a change in energy.

Your brain records audio, visual, tactile, and other types of information through the seven senses, which are stored as effects. As you review these effects, you become aware of the passage of time. But how does time actually pass? The passage of time consists of events and incidents that occur due to variations in energy, and you experience the effects of these variations. Nothing more than that.

Time is the Energy Variation

The concept of time has fascinated scientists, philosophers, and physicists for thousands of years. If energy variation were to stop, there would be no time, motion, space, or even a universe—no you and no me. The definition of time, despite its simplicity, occupied my thoughts for many years. This is because it raises numerous other questions, including the nature of the creator and the origin of the universe from nothing. We will explore these ideas through the lens of science and quantum physics.

Mathematical formula for time

The mathematical formula for time is the ratio of the input energy to the output energy. Therefore, any mass that absorbs energy A will return the same amount of energy.

T: Time

Ein: Energy input

Eout: Energy output

T=EinEout=Ein^2=Eout^2=E^2

If Ein=0, then T=Eout0=0 → T=0

If Eout=0, then T=Ein0=0 → T=0

If Ein=0 and Eout=0, then T=0

This formula indicates that if T equals zero, then there would be no universe.

Condition 1: Mass exists as a monopole when we lack the tools to identify monopole masses, and it is defined as nonexistent.

Condition 2: In reality, there is no mass. If the mass does exist, it must be confined to a specific range of time and will be influenced by changes in energy. It can be found within a state of nonexistence.

Condition 3: Mass exists within the realm of time, specifically in the present, represented as 1 ÷ ∞ # 0.

According to the twenty laws that govern the structure of the universe, as well as the law of the bipolar universe, all objects and particles within it are bipolar, consisting of both positive and negative charges. When two objects or particles are brought together, the positive charges from mass A exert a gravitational force on the negative charges from mass B, causing them to attract each other. This attraction continues until the repulsive forces—resulting from like charges—prevent the two masses from getting any closer by repelling particles away from each other. Thus, energy exchange occurs, and time is created. We will explore this concept further later.

From this discussion, we conclude that "time zero" is unique to the Creator, who exists outside of physical matter. We will explore the idea that the Creator is an intelligent being who brings time into existence yet is not bound by it in a material sense.

PART 2: PLACE

Place is the environment in which we perceive, feel, and inhabit. At the beginning of the universe, time and space came together, forming the fundamental foundations of existence. They are intrinsically linked; if time exists, space also exists, and vice versa. Although time and space emerged simultaneously, it is believed that an initial mass with infinitely high density and bipolarity was created in the form of an infinite supercloud of atoms. I have named this infinite cloud of atoms "Tanerium." In Turkish, Tanerium means "God."

According to the law of similarities—specifically, Law Number 8 of the 20 rules governing the structure of the universe—Tanerium comprises a cloud of atoms organized into seven electron layers and possesses a positive gravitational charge. Additionally, in accordance with the law of the bipolar world, a layered structure formed around the mass of Tanerium, fulfilling the principle of positive and negative balance within a single entity. This surrounding layer has a negative gravitational charge.

It is important to note that the nucleus of the Tanerium atom cloud has a positive charge. The first layer surrounding the nucleus has a negative charge, the second layer has a positive charge, and the third layer has a negative charge. This pattern continues until the seventh layer, which also has a negative charge and is located near the outer cloud. I refer to this as the "Paravazium" layer.

These two layers of the same name began to repel each other. The Paravazium layer, which leads to a state of nothingness (defined as a unipolar universe, meaning it has neither a positive nor a negative charge), began to retreat at a speed of millions of kilometers per second. The sixth layer of Tanerium, possessing a positive charge, is drawn toward the seventh layer due to the negative gravitational force of the Paravazium. Meanwhile, the fifth layer, which carries a negative charge, is pulled towards the sixth layer. This pattern of attraction continues for the subsequent layers.

As the Tanerium mass expands, the universe has grown to its current size.

The universe was created from nothingness. The cloud of the Paravazium layer has begun to retreat, and its current retreat speed has now reached 300,000 kilometers per second. The reason why light travels through space in a spherical form and why its speed remains constant is due to the gravitational force of the Paravazium layer, which scientists refer to as "dark energy." The speed at which energy varies and the flow of time are directly related to the speed of the retreat and the gravity of the Paravazium layer. If the Paravazium layer were to stop accelerating or if its speed were to reach zero, the universe would revert to nothingness. Such a day may come when the speed of light reaches zero, causing the universe to disappear.

The universe exists within nothingness. To illustrate this, think of nothingness as water and the universe as a sponge ball; just as water permeates and surrounds the sponge, the universe moves through nothingness. When the mass of Tanerium with its Paravazium layer was created, tiny, colorless spherical particles emerged and filled the space. I named these particles "Alium," considering them one of the greatest masterpieces of creation. These spherical particles are responsible for making images appear smaller at a distance. It has not occurred to physicists and scientists to explore why images seem smaller the farther away they are. Throughout history, humans have developed various devices, such as cameras, telescopes, and microscopes, to address this phenomenon.

PART 3: THE BIPOLAR WORLD

The first thing that captures human attention in the world is the concept of bipolarity. Examples of this can be seen in many contrasting elements: night and day, cold and heat, darkness and light, ugliness and beauty, low and high, male and female, attraction and repulsion, and many others. This notion of opposites is a fundamental aspect of our universe .

Even a compass has two poles: north and south. This bipolar system is present throughout the cosmos and can also be observed in the cells of living organisms, including both plants and animals. Although we may not often recognize it, this principle is a significant aspect of our reality.

All particles and objects in the universe possess positive and negative charges within their structure. The layer closest to the nucleus carries a negative charge, followed by a positive charge in the next layer, then negative again, and so on, alternating until the seventh layer. This alternating charge pattern is designed to ensure that atoms, along with their electron layers, remain stable and do not disintegrate. This structure mirrors the organization of the universe itself, which is currently expanding. The universe operates according to scientifically established laws, and we have a body of knowledge regarding its origin. Some individuals believe that the universe came into existence by chance or accident, which is quite perplexing. The question remains: How and where did this supposed accident occur? These beliefs are often vague and lacking in clarity. Respected guests, it is essential to note that the universe was formed based on scientific principles. We will communicate these laws as they are grounded in science. If there is any ambiguity in certain areas, it may be due to the limits of our current understanding. I refrain from expressing opinions without a solid scientific basis or a clear purpose.

PART 4: THE FIRST AND LAST LAW

This law states that everything has a beginning and an end. We can observe this process in all aspects of life, from plants and animals to the cycles of night and day, the seasons of the year, and the movement of the moon around the Earth, as well as the planets of the solar system around the Sun. The universe itself also has a beginning and will eventually face an end; however, this destruction will not be permanent. The universe will be created anew.

In the creation system of the super-intelligent Creator, no evolving plan or program seeks a newer and better method. What the Creator has chosen is the best and will remain the same forever. Perhaps I have written this book trillions of times, and you have read it just as many times; this cycle will continue indefinitely. Everything subject to time will come into being and eventually perish. This repetition will go on as long as the universe exists.

Only the Creator is eternal, having neither a beginning nor an end. Being non-material, He is the origin of time and is not constrained by the first and last moments. He exists solely in the present, where the past and future are meaningless to Him.

PART 5: TANERIUM MASS

The first mass that emerged from nothingness came into existence in pairs, entered the time domain, and had an infinitely high density. This mass is referred to as the "Tanerium super atomic mass." Under the influence of the gravitational force of the Paravazium superlayer, this mass expanded, which is why we observe the vast size of the universe today. It is essential to note that the universe is continually growing, not contracting. According to the 20 laws governing the structure of the universe, it is in a constant state of expansion. The idea that the universe is contracting or that particles are accumulating after the Big Bang—leading to the formation of stars and planets—is not acceptable. If that theory were accurate, all stars, planets, galaxies, and the entirety of the universe would eventually come together to form a single star, which is not happening. In my opinion, scientists may not fully support this theory because they do not adequately consider the driving forces at play.

Physicists identify four fundamental forces in nature: 1. The electromagnetic force, 2. Weak nuclear force, 3. Strong nuclear force 4. Gravitational force.

Surprisingly, physicists of the 20th and 21st centuries have overlooked mainly the concept of repulsive forces. They seem unaware that like charges repel each other and that similar repulsive forces exist between planets and stars, preventing planets from falling into stars. The law of the bipolar universe clearly demonstrates the existence of repulsive forces in contrast to gravitational forces. In fact, gravitational forces cannot exist in nature without the presence of repulsive forces .

First, the universe did not originate from an explosion, contrary to popular belief. Second, when we discuss the initial blast of the Big Bang, it suggests that the universe emerged randomly and without purpose, which is not the case. We must seek the truth based on the laws governing the structure of the universe. Initially, a mass referred to as "Tanerium" was created, forming a superatom with trillions of superatoms in its electron layers. After the universe expanded, these layers became increasingly dispersed, and each of those superatoms evolved into a galaxy.

Law number eight confirms the theory that the universe was created in the form of an atom with seven layers. Additionally, Law Number Seven supports this idea by stating that the universe functions as a whole; for example, the solar system and the Milky Way galaxy are both subsets of the universe. Each subset within a more significant subset is composed of even smaller subsets. The universe exists with remarkable order and precision. Law twelve highlights the monotheistic nature of the theory, positing that the universe has a central core composed of millions of subsets, each with its own spherical core. This spherical shape is effective in maintaining order by effectively controlling and containing its components through the force it exerts.

The universe has a central core, within which time is considered to be zero. This is because time in the universe originates from the super-force layer known as Paravazium. At the central core, the resultant gravitational forces balance out to zero. From this perspective, the central core of the universe can be viewed as a hollow sphere that eliminates energy variations that have settled at zero. Consequently, ideas and theories that are not in harmony with the natural order are ultimately worthless.

Some ideas and theories proposed by 20th-century physicists are no longer applicable, as today's physicists recognize that Newton's dynamics and Einstein's theory of relativity cannot fully explain questions about the universe's structure in the realm of quantum physics. Instead, we must seek a scientific understanding based on the 20 fundamental laws governing the universe's structure. Physics is my favorite science; it explores the universe's structure and composition. Therefore, I decided to share the information I have collected over several decades with the world so that others can benefit from it.

The content of this book is quite profound, but I have made every effort to ground it in scientific principles. It may take some time for both the public and physicists to grasp the ideas presented here fully. However, I have complete confidence in their validity and believe that a significant revolution is on the horizon for quantum physics. Although I am not a physicist, my research in the philosophy of creation has led me to explore quantum physics, uncovering mysterious truths through both scientific and philosophical means. This endeavor has been one of my most significant research aspirations.

Humanity has moved beyond classical physics and must now seriously engage with quantum physics to address the challenges we face today and in the future. Growing populations, diminishing food resources, and inadequate welfare systems, combined with self-inflicted challenges stemming from ignorance, threaten not only our existence but also that of other living beings, pushing us closer to decline. However, this generation's ignorance is fading as future generations increasingly reject such thinking. The advancement of human societies relies on the progress of knowledge and the betterment of humanity. As science progresses, it shapes the development of both. Unfortunately, uninformed scientists, politicians, and capitalists are significant obstacles to this progress. It is disheartening to witness philosophy and logic being undermined by their actions.

PART 6: MODERATION IN THE TENDENCY TO EXTINCTION

The Sun has been shining in our solar system for billions of years, with planets orbiting around it. Currently, the distance between the Earth and the Sun is approximately 150 million kilometers. According to the 16th law of the expansion of the universe, this distance has undoubtedly been much smaller over billions of years, during which the Earth orbited the Sun in a molten state. As the Earth's distance from the Sun increased, it gradually cooled to its current temperature, allowing for the development of single-celled organisms, bacteria, plants, animals, and eventually humans.

While we do not know how many times various species, including humans, have appeared and disappeared from the planet, evidence suggests that life has existed on Earth for at least a billion years. The changes occurring in the solar system are so gradual that they cannot be detected without modern technology and equipment. However, when we observe the sky with advanced telescopes, we often witness sudden and unpredictable celestial events.

Evidence suggests that the universe follows a specific order. Yet, current changes indicate that this order is not eternal and is gradually deteriorating. Although our lifespan is too short to observe these changes directly, research consistently shows that the universe is undergoing a decline.

When the Earth formed from a molten state filled with heavy atoms, an explosion gave birth to it and other planets. However, planets' separation from the Sun occurred over millions of years, not simultaneously.

As the Earth's crust cooled, it enabled the creation of oxygen and hydrogen, which subsequently formed water and air—processes that continue to this day. Other planets also formed from explosions, but their separation from the Sun was a gradual process that took millions of years. The cooling of the Earth's crust, influenced by superheavy atoms, was crucial in producing oxygen, hydrogen, water, and air, and these processes persist.

Super heavy is a specific category of both heavy and light atoms, which we will explore further in subsequent sections. The weather and air on Earth consist solely of superheavy atoms. It is misguided for scientists to search for water on other planets. The principle of moderation in annihilation contradicts the law of conservation of energy, as the universe has evolved from wholeness to fragmentation, ultimately resulting in photons and Aliums that are absorbed by the Paravazium layer and vanish over time.

Even when the time comes to an end, the universe itself ceases to exist.

PART 7: THE LAW OF SUBORDINATION

The law of subordination is a universal principle that states that no particles or objects in nature are entirely independent; instead, they are influenced by greater forces and can exert control over other forces. For example, the solar system consists of several planets and moons, while the solar system itself is part of a more extensive system known as the Milky Way Galaxy, which is encompassed by an even larger entity called the Tanerium Universe.

The universe did not come into existence due to an explosion; instead, it originated from a state known as a super atom, which was initially at an infinitely high density called the mass of Tanerium. This mass consisted of seven electron shells, with super atoms orbiting around a central nucleus. Over time, each of these superatoms evolved into a galaxy. Superatoms were formed from smaller subsets in which superheavy atoms orbited their electron shells. As these superatoms gradually opened up trillions of times, they gave rise to lighter and lighter atoms. This process ultimately led to the creation of modern metals, including iron, bismuth, uranium, nickel, aluminum, and copper, as well as essential elements such as oxygen and hydrogen, which contributed to the formation of air and water.

Super atoms have an incredibly high density; just a spoonful could weigh as much as the moon. The stars that scientists refer to as neutron stars, which possess even higher densities, can actually be described as "super atomic stars." However, it's important to note that neutron stars do not exist in the universe. Firstly, no neutral particle called a neutron is found in nature. Secondly, even if a neutron star were to exist, it would not be able to produce an electromagnetic field. Only super-atomic stars and those with super-heavy and ultra-heavy atoms can generate an electromagnetic field.

PART 8: THE LAW OF SIMILARITIES

This law posits that the components of the universe are formed in similar patterns. For instance, the shape of all objects—ranging from electrons and atoms to moons, planets, stars, galaxies, and even the universe itself—tends to be spherical. This is primarily due to the 12th law of centrality, which states that in a spherical state, the gravitational force of the nucleus can effectively maintain control over the objects within its influence. The gravitational fields of these objects extend into space in a spherical manner.

While plants and animals exhibit differences—mainly due to genetic variations influenced by climatic conditions—they still share fundamental similarities in their structure and design.

The genes of plants and animals are designed by an intelligent network within the universe's system of creation. The environment and climatic conditions determine these designs, as well as the elements present in each specific region. While all objects and particles possess a form of intelligence, this intelligence is limited to fulfilling their basic needs. They lack the capacity for design and complex thought. This topic will be explored in greater detail in the following sections. The laws governing the structure of the universe are interconnected, much like the threads of a fabric, and we are compelled to adhere to them.

PART 9: SYMMETRY SYSTEM

The symmetry system closely resembles the bipolar world system. Still, it also applies to the structure of objects, components, living beings, and plants, as well as their processes of emergence and eventual disappearance. The universe was initially created as a unified whole and then transformed into smaller particles before gradually fading away. This law fundamentally rejects the principle of conservation of energy. While the particles of the universe are the primary source of energy production, once these particles are transformed into energy, they do not revert to their original form. In essence, they are not destroyed. The symmetry system introduces a new principle of creation and destruction.

In classical physics, when mechanical energy is converted into electrical energy, the amount of energy obtained is not equal to the initial energy but is somewhat less than it. If you do this repeatedly, your energy will eventually dissipate to zero. In the symmetry system, the properties of the initial behavior of creation and the second behavior of destruction are opposite to each other. For example, the sun and the Earth have a gravitational force that pulls them towards each other. This distance reaches such an extent that they are attracted to each other. At this moment, the repulsive forces come into play. On the contrary, the gravitational forces prevent them from approaching and sticking together. This law is valid in all components and objects of the universe.

The components of the universe have positive and negative charges. The positive charges from the sun become homogeneous and, as a group, attract the negative charges of the Earth towards themselves. Conversely, the positive charges of the Earth attract the negative charges of the sun towards themselves, which causes the distance to decrease. As the distance between the sun and the Earth decreases, the charges of the same sign repel each other. This is the repulsive force. The distance between the Sun and the Earth, which is approximately 150 million kilometers, represents a balance between the forces of gravity and repulsion. Consider a magnet that has two poles, N and S, but there is no reaction in the middle. There is also an orbit between the Sun and the Earth, known as the Lagrange orbit. At that point, the forces of gravity and repulsion have reached an agreement.

PART 10: PAIRED CHARGES

This section explores the pairing of objects and particles. Quantum physics studies the fundamental structure of the universe. The universe was created according to the principles of science that we are now beginning to understand, and our progress relies on this knowledge. Classical physics, in general, does not explain how the universe came into being; these answers lie within quantum physics and particle physics.

The topic of paired charges is related to the concept of a bipolar universe. However, this section emphasizes that no particle or mass can exist as a monopole within the universe. For instance, if a mass is composed entirely of positrons, which have positive charges, the repulsive forces among the positive charges will cause the mass to disintegrate. Similarly, if it consists solely of negative charges, disintegration will occur again due to repulsive forces, resulting in destruction.

This principle teaches us that a mass or particle can only be stable if it contains both positive and negative charges. The most viable scenario is one where there are slightly more positive charges than negative ones within a mass. This indicates that in nature, we observe objects and particles with both positive and negative charges rather than those that are purely positive or negative.

In nature, we do not find monopolar or uncharged objects and particles. Physicists explain that while the proton has a positive charge, the neutron is often described as having a neutral charge. However, this neutrality conflicts with the fundamental laws governing the universe's structure, which are based on the forces of attraction between positive and negative charges.

How can a particle with no charge, either positive or negative, be accommodated within this framework? A neutral particle, like the neutron, would have neither gravitational attraction nor repulsive force. This leads to questions about how such a particle could exist within the atomic nucleus. It seems improbable that the atomic nucleus could accommodate a particle without a charge.

Thus, the neutron must possess some form of charge, potentially a negative one, to ensure the stability of the atomic nucleus alongside the positively charged proton. Additionally, a proton cannot exist in the form of a positron due to its tendency to decay.

Furthermore, the presence of a magnetic field arises because the atomic nucleus behaves like a magnet with two poles—positive and negative. The electron shells surrounding the nucleus act like coils that are constantly rotating, generating a magnetic field. If the nucleus were monopolar, a magnetic field would not form.

Ultimately, the atomic nucleus likely consists of both positive and negative charges, with the neutron opposing the positive charge of the proton by carrying a negative charge.

Law of paired charges states that it is inevitable for male cells to be present in the bodies of women and for female cells to be present in the bodies of men. This principle also applies to plants, suggesting that both male and female cells coexist in various organisms.

PART 11: THE INTELLIGENT SYSTEM

The Intelligent System We live in an intelligent world. Not only are we intelligent beings ourselves, capable of experiencing emotions, analyzing situations, and addressing our own needs as well as those of others, but we also manage our lives in increasingly complex ways. From the Paleolithic era to the present, we have made significant progress, enabling us to explore other worlds in the sky. In the distant past, humans used leaves to cover themselves; now, we inhabit towering skyscrapers. We can travel thousands of kilometers in just a few hours by flying. This showcases our intelligence.

Have you ever wondered where this intelligence originates? You might say it is due to the structure of your brain, but how did this brain structure evolve to give us such capabilities?

We will analyze this project from the perspective of quantum physics. The human brain, along with all living things, is composed of particles that form cells, which in turn are structured by molecules and atoms. Within the universe's particle structure, we encounter familiar elements, such as electrons and atoms. When a negatively charged electron is separated from its nucleus, it seeks another positively charged atom to pair with, thus completing itself. Similarly, when an atom loses an electron, it looks for another electron to regain balance. For instance, two oxygen atoms combine to complete their structure, or iron metal interacts with oxygen to form a stable molecule.

Have you ever wondered why these phenomena occur? You might argue that they arise from chemical properties, but there's more to it. I contend that this behavior reflects a sense of need within these particles. The chemical properties observed in solids suggest an inherent need. The electrons within atoms need to satisfy this requirement. When they do, they achieve a stable state. This sense of need serves as evidence of the particles' intelligence. If the particles lacked intelligence, they would not possess the inclination to develop these chemical properties. All particles, whether positively or negatively charged, have their own needs.

Every charged particle in nature has a need; no charge, whether positive or negative, exists without it. A charged particle inherently possesses a need, and the absence of such a need is impossible. In the universe, we do not find any particle devoid of charge or intelligence.

Humans do not experience needs when they are asleep or unconscious because they are not in a state of awareness. However, as soon as you wake up or become conscious, you begin to experience sensations such as hunger, thirst, tiredness, or relaxation. This awareness signals that you are in a state of intelligence. If you lacked intelligence, how would you even recognize the feeling of thirst and the need to drink water?

This need is communicated to your brain by the cells in your body, which indicates that your cells possess intelligence. They inform you when you need water. There have been times when you felt extreme tiredness or pain in a part of your body. Why is that? Fatigue is communicated through your cells, alerting you that harmful substances have accumulated in your body due to overexertion, and you need to rest to allow your body to detoxify. Similarly, suppose you experience pain in any area. In that case, it is your body's cells signaling for your attention so you can address the issue.

All parts of your body are intelligent, comprising a collection of billions of small and large intelligent particles. For instance, consider your liver, which produces hundreds of different enzymes. If you were to establish a petrochemical factory to replicate the functions of your liver, you would require several hundred hectares of land. Do you consciously direct your liver to function, or are you unaware of its role?

This discussion highlights that your body is composed of billions of intelligent cells that communicate their needs to you in various ways. They perform their tasks without your conscious knowledge, which serves as evidence of their intelligence. While you may be unaware of the specific workings of your organs, they operate with great care and intelligence.

Every particle in the world that experiences a need is considered intelligent. The need shows intelligence. All objects and particles consist of two opposing poles: negative and positive. The presence of these poles is the primary reason for the need.

Plants also possess intelligence. They are composed of subsets of needy particles and function as a more significant subset created to fulfill specific missions and tasks. Each plant is designed to meet particular needs and continues to perform its functions. This design is attributed to a super-intelligent creator of nature, responsible for the creation of the world.

The intelligent network of existence possesses comprehensive information about all the details of life. It has produced various plants according to geographical location, climate, and the animals and living beings in that area. When you plant a seed in the ground, and if the weather conditions are favorable, the plant will grow, bear fruit, or develop into a sturdy tree. This intelligent seed not only produces fruit but also supplies the oxygen necessary for living beings to breathe.

Many ornamental plants exist to enhance the beauty of nature by providing flowers and adding fragrance to the environment. I encountered a flower that was wilting due to heat and lack of water; it had bloomed prematurely. When I consulted an expert, he explained that the plant was dying. However, it was still flowering and producing seeds to ensure its reproduction for the following year, thereby providing the survival of its species. It's remarkable how intelligent plants can be.

Many intelligent plants and animals on Earth are truly amazing. Take, for example, the life of a honeybee. It exhibits a remarkable organization and division of labor, having adapted to the Earth's climate over millions of years. If you observe ants, you will notice a complex form of intelligence in their behavior as well.

Many birds and animals can sense the changing seasons without the need for a calendar. They are able to find their way during migrations, knowing precisely where to go during the cold and hot seasons, and they return to their original locations. They gather in groups and migrate together without the use of loudspeakers, media, or compasses.

Although humans are intelligent, there are instances where certain animals in the wild exhibit superior intelligence. According to the third law of the bipolar world, one of the 20 laws of the universe, all particles possess both positive and negative poles; this principle is also found in quantum physics.

From this, we can conclude that the entire universe is composed of intelligence. If intelligence were removed from objects, nothing would remain—there would be no objects and no universe. The life of the universe wholly depends on intelligence, and ultimately, intelligence is all that exists.

Intelligence has evolved from a state of nothingness to our current world, as existence itself emerged from this primordial state of being. In this context, both nothingness and the Creator can be understood as forms of intelligence. While we cannot directly access nothingness, reason and logic indicate that we are subject to the passage of time. We believe that everything has an origin, and we consider nothingness to be that origin. Additionally, we know that the universe will eventually cease to exist, returning to nothingness, as described by Law No. 4, which governs the universe's structure. Thus, not only is matter and mass considered to possess intelligence, but intelligence itself is also a form of matter that will ultimately revert to nothingness. Undoubtedly, nothingness represents a type of intelligence and can be nothing else.

In quantum physics, there is a concept distinct from that in classical physics. When two masses, +A and -A, combine, they effectively cancel each other out, resulting in a net mass of zero. This raises an interesting question: What about intelligence? I believe the answer lies in the nature of intelligence itself, which is akin to nothingness. I assert with confidence that nothingness reflects the essence of intelligence.

$(+A) + (-A) = 0$

$(+\infty) + (-\infty) = 0$

When two masses, +A and -A, combine, they form a particle with no charge, making it invisible and unrecognizable to us. We typically identify particles by their opposite charges. For instance, we recognize darkness by the presence of light, cold by the presence of heat, and attraction by the absence of repulsion. Monopolar matter or mass lacks characteristics that we can feel or perceive. A monopolar particle does not occupy a place in the time domain; instead, it is associated with nothingness.

The intelligent network of nothingness has created the universe using a unique formula. We know that combining infinite positive and infinite negative results in zero, symbolizing nothingness expressed mathematically as $(+\infty) + (-\infty) = 0$. However, philosophy argues that zero inherently contains both infinite positive and negative aspects, which cannot be separated. In contrast, the intelligent network of nothingness has distinctly differentiated between the two, leading to the emergence of a bipolar world.

In addition to the Tanerium world, an infinite number of other worlds exist in non-existence, making our Tanerium world seem like a pebble in an endless ocean. These statements are grounded in science, suggesting that there is no reason to believe that our world is the only one; instead, there are countless other worlds.

The physical existence and the concepts of being and non-being encompass all of existence: humans, animals, plants, matter, and non-matter. This existence originates from a fundamental type of intelligence—the same intelligence that creates everything from nothingness, destroys it, and then recreates it.

In this book, I explore the 20 laws that define the universe's structure from a philosophical standpoint, drawing on years of research in quantum physics. These laws are interwoven like threads in a carpet, mutually reinforcing one another. I aim for both the general public and scientists to grasp and benefit from these insights.

It is essential to recognize that the universe was created from nothingness and intelligence. Contrary to popular belief, the universe did not arise by dismissing these elements; instead, existence is encompassed by both nothingness and intelligence. The realm of reality cannot function apart from the influence of intelligence. Just as a fish resides in the ocean's depths, our world exists within the realms of intelligence and nothingness, with nothingness permeating all its particles.

The time-dominant universe, composed of bipolar components, is influenced by a supergravity force known as the Paravazium force. This force prevents these components from merging, as their union would lead to nothingness. Nothingness signifies a unipolar intelligence that can exist apart from immaterial components. These components pervade both the Tanerium universe and the realm of nothingness, and there are no other scenarios.

I argue that the essence of all existence and nothingness comprises identical, separable components known as the Creator or God. Concerning the universe's formation, it was the intelligent super-infinity of nothingness that converted the initial mass of the super-atom of Tanerium from separable, unipolar components into a bipolar structure, thereby creating the universe.

PART 12: THE MONOTHEISTIC SYSTEM AND CENTRALITY (DARK MATTER)

According to Law No. 7 of the Subset in the Laws of the Structure of the Universe, each subset is formed due to a strong central force that seeks to expand its sphere of influence. This central force dominates several other subsets through the force of gravity, ultimately forming a larger subset while remaining under the influence of an even larger subset. Without this central force, no subset would ever come into being. For instance, in the solar system, the Sun acts as the powerful central force that gathers several planets, each with their respective moons, thus forming the solar system as a subset.

In the Milky Way galaxy, the center consists of numerous stars, with a significantly more powerful star at its core, which governs the structure and dominance of the galaxy. Similarly, within an atom, the nucleus serves as the central core. For example, Jupiter has 67 moons orbiting it, while Saturn has 62 moons, all of which are centered on their respective planets.

There exists a centrality in the universe. Although the universe originated in the form of an atom, its central core has the power to control seven celestial layers, encompassing all galaxies and universes. While the universe is centralized in terms of materiality and mass, it cannot be centered on metaphysical intelligence. The universe was born from intelligence, which spread throughout it, making it impossible for there to be a singular center of origin.

Consider any point in the universe; that particle serves as the center of the entire intelligence within the universe of nothingness and the current universe. At the heart of each galaxy lies a star with significantly higher mass and density. When a star's life comes to an end, it transforms into a massive black hole, which eventually disappears after millions of years. As this occurs, the galaxy becomes decentralized, collapsing under the influence of gravitational forces from Paravazium (hidden energy). Consequently, the galaxy loses its spherical shape. Numerous galaxies exhibit this behavior, indicating a potential end to the universe.

PART 13: SUPER FORCE OF PARAVAZIUM (HIDDEN ENERGY)

According to Law No. 3 of the bipolar universe, all particles in the universe are created as bipolar entities. In my view, the central core of the Sun carries a positive charge, whereas Mercury has a negative charge. Similarly, Venus has a positive charge, Earth has a negative charge, and this pattern continues for the other planets in the solar system.

The first mass in the universe, known as the superatom of Tanerium, had a positively charged nucleus with seven layers, although the seventh layer carried a negative charge. Overall, the superatom was positively charged, akin to the sun in our solar system. To achieve balance, a negative charge was needed to counteract the nucleus's positive charge, enveloping the superatom. The outer layer was called Paravazium.

Tanerium has an infinitely high density. In its electron shells, Tanerium mass orbits around the nucleus instead of the electrons found in superatoms. The number of these superatoms is infinitely large, with each one potentially evolving into new stars and systems as the universe expands. The orbit of the electron shells around the nucleus generates the largest magnetic field in the universe.

As the Paravazium super-force layer rotated irregularly around Tanerium, its negative charge interacted with the negative charge of the seventh layer, causing them to repel each other. Consequently, the Paravazium super-layer found itself in contact with nothingness. This nothingness lacks material substance and is characterized by intelligence.

Nothingness is unipolar and can be described as a quantum vacuum devoid of matter. In this quantum vacuum, the Paravazium layer initially began to retreat at a speed of millions of kilometers per second. It is essential to clarify that by "nothingness," I mean a state without the dominance of time or any matter within the quantum vacuum.

At the moment of creation, the mass of Tanerium and the Paravazium superlayer were formed simultaneously, along with the Aliums. Aliums, governed by law number 17 of the universe's structure, are infinitely small particles that take the shape of colorless spheres. These particles filled the entirety of space and the universe. In Section 17, we will discuss these spherical particles, known as Aliums, in detail.

Aliums are particles that propel atoms with positive and negative charges toward one another, thereby creating space. They are not attracted to any natural objects. Aliums fill all the gaps in the electron shells of atoms and molecules throughout the universe. Without Aliums, radio frequencies of light and magnetic fields would be unable to propagate through space, and spacecraft would struggle to move in space without the frictional force provided by Aliums.

Aliums played a significant role in filling the vacuum and expanding the universe. Although the Paravazium layer initially retreated at a much higher speed, this rate is now decreasing. It has now reached a speed of 300,000 kilometers per second. Eventually, there will come a time when the speed of light reaches zero, resulting in the universe's eventual disappearance. I estimate this will occur in less than 6 billion years.

The primary reason for the movement of all objects in the universe is the gravitational force of the Paravazium layer. This gravitational force is the main factor that causes planets to rotate around stars and governs the systems within the galaxy.

Why do planets orbit their stars, and why don't planets fall into stars?

According to the 20th law of the universe's structure, nothing can be at rest. This is because all objects and particles possess both positive and negative charges and are subject to gravitational forces and thrust from all directions.

The primary reason planets orbit stars is the gravitational force generated by the mutual attraction between the stars and their planets. For example, the Sun exerts a strong gravitational force on the Earth's negative charges due to its positive charges. Likewise, the Sun's negative charges also exert a powerful gravitational force on the Earth's positive charges, causing the Sun to pull the planets toward it.

In addition to the gravitational forces from the Sun and the Earth, the Sun's gravitational attraction counteracts the Earth's gravitational attraction. In contrast, the Sun's negative thrust forces correspond with the Earth's negative thrust forces. This interaction helps maintain a specific distance between the two bodies, preventing them from moving closer to a certain point.

Surprisingly, physicists have focused so much on gravitational forces without addressing thrust forces. This relationship is crucial for explaining why planets are attracted to the Sun but do not fall into it. This law also applies to the structure of atoms.

According to the 10th Law of Paired Charges, all objects and particles in the universe have both positive and negative charges. Without repulsive forces, objects, planets, stars, and even galaxies would be drawn toward one another. Scientists believe that centrifugal force prevents planets from colliding with stars; however, the 20th Law of the Structure of the Universe completely dismisses this belief.

The question is: Where and how is centrifugal force produced? This phenomenon can only be fully understood through quantum physics, not classical physics. One key reason planets do not fall into stars across all galaxies is the presence of repulsive forces.

Many scientists hold an overly simplistic view, believing that stars and planets form only through the accumulation of objects and particles under the influence of gravity. However, there are several points to consider:

According to the sixth law, the universe is expanding, not contracting. In addition to gravitational forces, all objects possess repulsive forces that prevent them from coming together and forming large masses. Thirdly, according to the seventh law in the subset, objects are created as subsets, which expand, not collapse.

Finally, the law of moderation regarding the tendency toward destruction suggests that the universe has an underlying order; however, within this order, everything is moving toward destruction, with particles disappearing rather than being absorbed or merging into a single entity.

Why do planets and stars spin on their axes?

We have discussed the distances between planets and stars sufficiently. According to the 20th law of the universe's structure, no mass or particle in the universe can remain stationary. Alongside the gravitational and repulsive forces acting on particles and other particles, objects and particles are also influenced by a phenomenon known as the Paravazium superlayer.

The Paravazium superlayer rotates around the universe, exerting gravitational forces that pull all particles, as well as planets and stars, toward itself, causing them to move. However, these objects are also affected by the gravitational forces of their stars, meaning planets orbit around stars. At the same time, all galaxies, universes, and systems are also in a state of rotation around their centers.

If there were no Paravazium layer, objects would remain at a fixed distance from one another and would be unable to move. Over billions of years, all the planets in the solar system have gradually separated from the sun and continued to drift away from it, with some reaching the end of their life cycles. Currently, Earth enjoys a favorable climate that supports the existence of living beings. The gravitational forces of the planets have weakened over time as they have moved farther from the sun. This movement, influenced by the momentum and gravity of the Paravazium, has caused the planets to drift further from the sun each day.

PART 14: THE LAW OF REINCARNATION

According to the law of reincarnation, nothing in the world is ever truly destroyed; instead, what has been created will eventually be dismantled and subsequently recreated. This process will continue for as long as the universe exists, governing the cycles of day and night, the changing seasons, and the growth of plants, animals, and even humans. The universe itself, having been created, will also undergo destruction and be renewed.

Through this ongoing cycle, you and I have existed and perished trillions of times, returning in the same form and appearance. The universe follows a singular, constant process. It's important to note that this statement does not apply to the principle of conservation of energy. When matter is converted into energy, it is effectively destroyed in its original form and cannot be reverted to matter.

However, humans are regarded as eternal beings. Due to our intelligence, we are thought to return to life in another body after death, a concept that aligns with principles found in quantum physics. When we refer to a living being or a human, we are speaking of its intelligent system. A living being without metaphysical intelligence is merely matter. Those who believe that intelligence is simply a property of matter are mistaken. While matter may demonstrate intelligence, it is restricted to fulfilling basic needs.

Humanity possesses the intellect, wisdom, and capacity for metaphysical analysis. This metaphysical understanding has shaped the universe, transforming it from a state of nothingness into one dominated by time and space. Scientists worldwide agree that the universe is on a path to destruction. The concept of time dominance is also facing collapse.

Earlier, I mentioned that the speed of light at the beginning of the universe was millions of kilometers per second. Today, it has stabilized at approximately

300,000 kilometers per second, and this speed is gradually decreasing. When it eventually reaches zero, the universe will face destruction, as the speed of change in energy or time is equivalent to the speed of light.

This destruction will not happen all at once; instead, it will occur gradually, piece by piece, until the universe is annihilated.

The primary reason for the decrease in the speed of light is the reduction in the number of positive charges. Positrons are diminishing rapidly, leaving only negative charges in their wake. The presence of these negative charges creates repulsive forces that repel one another. Consequently, the world could eventually turn into a unipolar powder, which, through contraction, might dissolve into nothingness.

Metaphysical intelligence in humans represents a form of consciousness related to the concept of nothingness while also acting as the creator of existence. This intelligence does not disappear but continually transforms. When we refer to "man," we mean the soul, not the physical body. The soul possesses the body and is in a constant state of change. Therefore, man does not really die.

PART 15: THE LAW OF NUCLEAR CHANGES

This law is one of the fundamental principles governing the universe's structure and plays a crucial role within it. Law number 4 of the universe's structure, known as Tanerium, states that everything that has a beginning will inevitably have an end. Meanwhile, Law number 16 regarding the expansion of the universe indicates that it is continuously expanding and growing.

This expansion happens because larger objects break down into smaller, denser particles with reduced volume. As these larger entities transform, they create smaller objects that have greater volume and lower density. Consequently, the overall volume of the universe increases while smaller objects are converted into particles and gradually vanish.

The law regarding number 7 of the subset indicates that the mass of the Tanerium superatom consists of a collection of 7 electron shells, within which trillions of superatoms orbit the nucleus. Following the expansion of the universe and the opening of the electron shells, each of these superatoms is composed of other superatoms. As the universe expands, this process suggests that larger subshells are breaking down into smaller subshells, which in turn are transforming into smaller particles of Alium.

These particles not only increase the overall volume of the universe but are also gradually drawn into a lower layer due to the gravitational pull of the Paravazium layer, ultimately disappearing.

The world is constantly changing because time is an ever-present force. Larger subatomic particles are destroyed and disintegrated, becoming heavier and then lighter atoms. Eventually, the lighter atoms in the outermost orbit lose an electron, transforming into smaller atoms, such as hydrogen and helium, which also naturally decay.

According to the law of subatomic particles, heavy atoms do not have electrons in their outer shells, while lighter atoms have electrons circulating in these shells. Only certain electrons in the outermost shell can be replaced. This could be a definitive and well-established theory.

According to the law of similarity, both a solar system and a galaxy can be compared to atoms. In this analogy, a solar system represents an atom, with multiple planets and moons orbiting other planets. A galaxy, on the other hand, consists of billions of stars, each with several planets orbiting it, and these planets often have their own moons.

In this context, a helium atom can accommodate two hydrogen atoms in its orbit. Generally, heavier atoms remain stable only at high temperatures. For instance, during a hydrogen explosion, the surrounding temperature increases, facilitating helium production. If the temperature becomes too high, helium can transform into a heavier element.

Additionally, according to the law of expansion, the universe is continuously expanding, and objects within it are gaining volumes over time.

PART 16: THE EXPANSION OF THE UNIVERSE

When the volumes of super-atomic Tanerium fused with the supergravity of the surrounding Paravazium, tiny spherical particles called Aliums formed. This process of expansion has continued since the beginning of creation. The Aliums filled the resulting vacuum we know as space. The universe is entirely composed of mass; nothing else exists apart from it. Aliums are colorless, infinitely small particles of matter that can act as powerful agents of force transfer.

PART 17: ALIUMS

For many years, I wondered why images appeared smaller from a distance. To find the answer, I referred to physics books and other sources, but I was surprised that no one else seemed to have considered this question. Why haven't physicists commented on it? After I began researching quantum physics, I found it easier to explore this question in greater depth.

One day, while contemplating the nature of the universe, I realized that humans invented cameras and telescopes to help us see distant objects, such as planets and moons, and microscopes to observe tiny creatures like microbes and viruses. This led me to wonder if there is something in nature that operates in opposition to these devices—perhaps spherical particles that reduce the size of images, allowing us to perceive them more easily. I decided to call these colorless spherical particles "Aliums"

Through further research in this area, I discovered that Aliums are one of nature's greatest masterpieces.

Aliums fill the entire space. Albert Einstein, the renowned scientist, believed that space is filled with photons. However, this belief is not entirely accurate, as photons do not constitute space; instead, they represent the frequencies of light and energy. Furthermore, photons can only move through space by means of Aliums. Without Aliums, photons cannot navigate through space because there is no other medium that provides friction within it.

For example, if you hold two strong magnets with like poles facing each other, you will find that they cannot be brought close together because their magnetic fields repel each other. Although the force is invisible, it prevents your hands from getting closer. On the other hand, if you bring opposite poles together, you will find it impossible to keep your hands apart, as they will attract and stick to each other.

This interaction demonstrates the presence of a magnetic field. But how is this field transmitted? The only means of transmitting this force are the Aliums, which neutralize the force exerted by your hands.

In addition, consider the air in the gap between the magnets. While air molecules do not possess a magnetic field, the spaces between electrons, atoms, and molecules are filled with Aliums. Without Aliums, all electrons would fall into the nucleus and merge into one entity, leading to the disappearance of stars, planets, and galaxies. Aliums serve as the transmitters of gravitational and thrust forces.

Outside of Earth, in the vast spaces between stars and planets, where atoms and molecules are virtually nonexistent or present in tiny quantities, it is the Aliums that effectively transmit magnetic and electromagnetic fields, both in a vacuum and on Earth. Without Aliums, there would be no electric current. If Aliums did not exist, humanity and the entire universe would lack a genuine existence. Even if we did exist, our eyes would need to be as large as trees in order to perceive them!

I have long wondered how spaceships fly in a vacuum without friction. After discovering Aliums, I concluded that they might be the only frictional force affecting the movement of spaceships. Aliums not only guide radio, optical, and electromagnetic frequencies in space, but they also shield us from harmful rays, including X-rays, alpha rays, beta rays, and gamma rays, that can reach our eyes and bodies.

Aliums are bipolar, meaning they have two opposite poles or charges. When placed next to each other, they exert pressure on the molecules in your hand, neutralizing the force of your grip. They play a significant role in electromagnetic fields, forming tiny particles and masses that flow in wires as electric currents. Interestingly, the electric current in cables is composed of Aliums, not electrons. We do not have free electrons in nature, and there is no inherent reason for their existence. One might wonder where free electrons come from, anyway. The only things that exist abundantly and freely in nature are the **aliums**. All magnetic and electromagnetic fields are made up of these elements.

It is truly amazing! What an infinitely intelligent creator has crafted the universe! He has created things that we humans struggle to comprehend. This discovery can only be explained through the lens of quantum physics.

Why do the sun and moon look larger when they are on the horizon? What is the primary reason that mirages and images appear larger in the desert?

It was previously discussed that Aliums, which are small and large particles that occupy the entire universe, serve not only as means of transmitting power and energy but also as tools that alter our perception of images at a distance. Aliums vary in weight and volume, taking on spherical and elliptical shapes. Those with greater volume and weight are found near the surface of the Earth, causing the moon and sun to appear larger on the horizon. Essentially, Aliums act as spherical mirrors.

Regarding mirages, it should be noted that Aliums transfer images of the sky to the surface of the Earth, creating the illusion of water for observers.

The reason images appear larger in the desert is due to the spherical particles in the hot, dry conditions, which become elongated and oval, with their apex facing downward. This distortion can lead to the perception of images being larger— such that you might see a person 3 kilometers away appear as though they are 50 meters tall. I have witnessed such optical illusions myself in the desert.

PART 18: MASS, WEIGHT, AND CHARGE IN OBJECTS

According to Law No. 3 of the Structure of the Universe, the universe is created in a bipolar manner. Law No. 18 regarding mass, weight, and charge in objects states that no mass or particle in the universe exists without a charge. All particles possess both positive and negative charges due to their gravitational force, which contributes to their weight. The presence of the gravitational force is the reason for the weight of objects and particles.

The renowned 20th-century scientist Albert Einstein claimed that photons, the smallest particles in the universe, do not possess a charge. However, this assertion is inaccurate. Photons consist of positive and negative light pulses. If they lacked weight or charge, they couldn't travel through space, be affected by the gravitational forces of planets and stars, or be deflected.

The main factor driving the movement of light through space is the gravitational force of the Paravazium layer. If a photon were to have no mass or charge, it would be puzzling how it could be affected by Paravazium gravity and traverse space. In that case, light deflection would be inexplicable.

How is light created?

Light exists in space in the form of frequencies and both positive and negative pulses. It is important to note that in nature, we do not find absolute positives or negatives; instead, we observe tendencies toward positivity and negativity.

Positively biased pulses are photon particles emitted by positively charged entities. In contrast, negatively biased pulses are photon particles emitted by negatively charged entities. These positively and negatively biased pulses can easily combine and spread through space as a spherical wave under the influence of the gravitational force of the Paravazium. Additionally, all objects and particles inherently possess a pole and weight.

Why is the speed of light 300,000 kilometers per second? Is the Speed of Light Constant?

Light is created by the resonance between like charges. The theory and definition of light have been discussed previously, and I will briefly summarize them here. Light consists of frequencies generated by the resonance between similar charges that are directed toward each other.

When two particles with positive and negative charges approach each other due to pressure, impact, or chemical changes, positive charges (comprising photon particles) are emitted from the positive charges of mass A toward the positive charges of mass B, resulting in a positive half-cycle or pulse. Conversely, negative photon particles are emitted from the negative charges of mass A toward the negative charges of mass B, creating a negative pulse.

These positive and negative pulses are generated sequentially, but they cannot move through space autonomously. Instead, they are influenced by the negative gravitational force of the super-Paravazium force, causing them to spread spherically through space. The Paravazium facilitates this spherical distribution of light frequencies within the realm of the Aliums.

The formula for the force produced by light is:

$$E = mc$$

This equation illustrates the relationship between mass (m) and energy (E), where mass is converted into energy when traveling at the speed of light, which is 300,000 kilometers per second. Thus, the mass of light particles (photons) can be transformed into energy.

The speed of light and the concept of time are not constant values and will not remain so. A significant number of galaxies observed through the James Webb Telescope have vanished, and we are merely seeing their light. According to Law No. 6 of the Law of Moderation in Tendency to Annihilation, no standard in the universe has ever been, nor will it ever be, constant.

We know that the universe's expansion speed at its inception reached millions of kilometers per second but is now decreasing. The speed of the Paravazium super force, which possesses a negative gravitational force and contributes to the universe's expansion, is the primary reason behind the current speed of light, which has stabilized at approximately 300,000 kilometers per second but is gradually decreasing—potentially reaching zero, leading to the universe's destruction.

The universe's increasing diameter (more than 28 billion light-years) and its estimated age (approximately 14 billion years) provide solid scientific evidence that both the expansion rate of the universe and the speed of light were significantly higher in the past. Although the James Webb Telescope has observed structures up to about 95 billion light-years away, we still lack precise information about the universe's exact diameter.

If we determine the exact diameter and age of the universe, we can accurately ascertain its past and future. I believe that the speed of light is decreasing at a rate of 5 centimeters per year, which suggests that the remaining lifespan of the universe will eventually reach around 6 billion years. However, the actual time may be much less than that.

If the universe's diameter is approximately twice 95 billion light-years, its remaining lifespan could be about 3 billion years. Assuming a diameter of 95 billion light-years would suggest a size roughly 3.4 times larger than the actual universe, which is 14 billion years old. This would require a light speed of 1,020,000 kilometers per second to achieve that diameter. However, the speed of light is actually 300,000 kilometers per second and is decreasing, varying in different regions.

A Challenge to Einstein's Relativity Formula

I believe that Albert Einstein's relativity formula, $E=mc^2$, has two main issues.

The first issue is that this formula is identical to the formula for light propagation. When mass is converted into energy during an explosion, it is influenced by the Paravazium layer. It disperses into space at the speed of light. Therefore, we can express this relationship as $E = mc$.

The second issue is that not all the mass of uranium or plutonium is converted into energy. Instead, lead and bismuth atoms remain in their electron shells and are not transformed into energy. Thus, the equation can be modified to:

$E = (m - m')\ c$, where m' represents the mass of the remaining lead and bismuth elements.

The principle of energy is one-directional: mass can be converted into energy, but not energy into mass. All natural objects emit undetectable energy that dissipates into the environment and cannot be reclaimed, effectively leading to their destruction. This destruction is evident in the world around us. The law of nuclear decay, as stated in the 20th law of the universe's structure, illustrates this phenomenon, as heavy atoms gradually transform into lighter ones until they vanish.

Given that no one has successfully converted energy into mass, why should we accept the equivalence principle? Using classical physics formulas, we find that during an explosion, mass "m" is converted into particles traveling at the speed of light, thus yielding the relationship $E = mc$.

In discussing the theory of probability in physics, I want to convey that nothing in the world is truly random or purely probable. Instead, what we call probabilities or coincidences arise from our lack of understanding of their scientific origins. Current science has not yet progressed to a point where it can thoroughly answer every question. The reality is that probability and coincidence do not inherently exist in the fabric of our existence; they are merely concepts we use because we are unaware of the underlying scientific principles. Every event that occurs has a scientific reason and foundation. Our perceptions of probabilities, incidents, and coincidences stem from our ignorance of their true causes.

PART 19: FORCES IN THE UNIVERSE

There are four fundamental forces in the universe:

1. The Paravazium Superforce or Hidden Energy: This force has been discussed extensively before.

2. Electromagnetic Forces

3. Gravitational Forces

4. Repulsive Forces

1. The mass of the superatom known as "Tanerium" was initially created with a positive tendency. Surrounding it, a mass of Tanerium with a negative tendency formed in a spherical shape. I have already explained how the Paravazium Superforce expanded the universe and shaped it into what it is today.

2. Electromagnetic forces are produced exclusively by superatoms and superheavy atoms. In this context, the nucleus of superatoms acts like a magnet. Surrounding the nucleus, numerous superheavy atoms rotate in a coil formation, generating a significantly stronger gravitational and repulsive force. This arrangement creates a robust electromagnetic field. As the number of these atoms increases and they come into phase with each other, they can travel millions of kilometers. This phenomenon also occurs in the sun, which is composed of super-heavy atoms.

3. Gravitational forces exist throughout nature and in all objects, particularly around stars and planets, causing them to form and remain cohesive. This gravitational force is also present in atomic nuclei, functioning similarly to the gravitational force that holds other elements in the solar system, which is composed of electrons.

4. Repulsive forces are the opposite of gravitational forces. All objects consist of positive and negative charges. When two objects are placed close together, the positive charges from mass A exert a force on the negative charges of mass B, and vice versa. When these objects are a certain distance apart, a strong opposition occurs from like charges on both sides. This phenomenon prevents the two masses from merging. Repulsive forces exist between all stars and planets, as well as galaxies and universes, stopping them from combining. The primary reason planets do not fall into stars is these repulsive forces, rather than centrifugal ones.

PART 20: NO MASS OR PARTICLE CAN STAY STATIONARY IN THE UNIVERSE

According to Law No. 18, all objects and particles possess positive and negative charges, have weight, and are in constant motion due to the influence of gravity and the forces exerted by other objects. Law No. 13 states that superparavazium has a negative gravitational force that attracts all stars, planets, galaxies, and universes toward itself while also imparting a rotational state to them. Law No. 10 states that all objects in nature possess paired charges and that both positive and negative gravitational forces exist within them. Importantly, we do not find complete positive charges, complete negative charges, or neutral charges in nature.

Any charged object or particle cannot remain at rest. In this context, we are examining the fields of quantum physics and particle physics. While objects on Earth may appear stationary, they are composed of tiny particles that are in constant motion. Over time, these particles can disappear and disintegrate.

Each particle in nature that is separated from its source acquires a charge proportional to its mass and weight. A charged particle cannot remain stationary in the presence of gravitational forces exerted by other particles or objects.

How do the gravity and thrust forces operate in the universe.? The force of gravity in Paravazium ,as well as the forces of gravity and thrust in space, are uniquely applied through Aliums. First I want to emphasize that all of space including the gaps between electron shells and nuclei, is filled with Aliums . These Aliums exist as infinitely small and colorless spheres with a central nucleus,which makes it challenging for them to escape their spherical state .In response to the two negative and positive charges of masses A and B ,these tiny spheres become egg-shaped and ,while trying to return to a spherical state, draw the two desired masses closer together. On the other hand ,it is essential to remember that different elements ,when faced with two negative or two positive charges ,take on shell-like or disk-like forms, which remains unacceptable for them .

They are compelled to revert to a spherical state , which prevents two celestial bodies from approaching one another .This interaction oscillates between gravitional forces and thrust forces until it reaches a state of equilibrium , positioning the two celestial bodies next to each other .

Why are the orbits of planets not perfectly circular?

Why are some planets completely locked while others are almost locked?

Two important reasons can be identified in response to this question. First, the presence of high mountains and elevations can influence the gravitational effect from the star, affecting the distance between the planet and the star and the angle between the planet's rotational axis and its poles.

The second significant reason is that all planets are separate from their stars. The interiors of stars contain superheavy atoms, and when a star explodes, molten material along with these superheavy atoms can be transported to the newly separated planet. Over time, the Earth's surface solidifies. However, at the moment of initial separation, the superheavy atoms are not evenly distributed throughout the planet; they tend to be slightly displaced toward the shell on the opposite side. This uneven distribution means that the planet is not perfectly balanced, causing the center of mass and the center of gravity to be misaligned.

As the planet orbits the star, the dominant gravitational force is concentrated on one side. This leads to two scenarios:

The first scenario occurs when the planet becomes entirely tidally locked, with its center of gravity consistently facing the central star due to its strong gravitational force. This situation is similar to how the Moon orbits Earth.

The second scenario involves the planet orbiting the star while still feeling the predominant gravitational force on one side. As a result, the planet rotates slowly, and its axis of rotation around the star becomes elliptical. Mercury exemplifies this situation.

It is rare for molten matter and superheavy atoms to be perfectly centered in a planet's core. This phenomenon also occurs on Earth, though infrequently. The Earth's molten core is not perfectly centered, which is why the magnetic pole is approximately 23 degrees off the Earth's axis. Over millions of years, planets gradually lose the superheavy atoms in their cores, causing them to emerge from their locked state and resulting in their orbits around the Sun changing from an elliptical shape to a circular one.

Gravitational field and its definition

The gravitational field is an area where gravitational force is exerted by a mass with both positive and negative charges through Aliums, attracting the opposite charges of another mass. The strength of the gravitational field increases with greater mass and decreases with distance. The force is projected in space only when there is a receiving source to respond to it; without such a source, the force does not propagate.

A thrust field is a region that exerts forces between two masses with the same charge as they approach one another. These forces, generated by Aliums, act on the opposing mass and prevent it from getting any closer. It is important to note that the magnitude of the gravitational forces at a distance is always greater than that of the thrust forces.

When wind or air hits your face, it isn't the electrons and nuclei of the air atoms impacting you; instead, they are Aliums located in the spaces between the electron shells and the nucleus. The Earth's gravitational forces act on an airplane through these Aliums, pulling it toward the planet. To counteract this, humans have developed engines that compress the atoms and molecules of air along with the Aliums, enabling the aircraft to overcome gravity.

When an apple falls from a tree to the ground, it does so not because there is no repulsive force between the two objects—there is one—but because its strength is so minimal that it cannot counteract the Earth's intense gravitational pull. In space, gravitational forces can extend for billions of kilometers. The moon's gravitational force, for example, causes billions of tons of water to be moved across the oceans during high tides. Still, buoyant forces keep that water from falling back to Earth.

This leads us to understand that nuclear gravitational forces exist within all atoms. However, strong electromagnetic forces are produced only by superheavy atoms because their electron shells contain a combination of heavy atoms with positive and negative charges orbiting the nucleus, creating a magnetic field.

Dead planets lack superheavy atoms in their cores, meaning they do not generate an electromagnetic field. As a result, such planets cannot have weather or an atmosphere, and they do not experience earthquakes or volcanic activity, nor will they in the future.

How did the universe originate from nothingness?

The universe did not emerge by chance; instead, it was created through a network of intelligence that existed within nothingness. Before the universe came into being, this intelligence already resided in the void. As humans, we perceive this intelligence in the universe, which aligns with the 11th law of intelligence. Although intelligence is not a physical entity, it can be observed through the presence of matter. It has been established that all objects and particles consist of positive and negative charges, which demonstrate a sense of need. Every need can be seen as evidence of intelligence. The foundation of intelligence stems from need, and every need originates from the interplay of positive and negative charges.

Life did not fall to Earth from elsewhere; it was created here because this planet is inherently ready for life in every way.

When two opposing forces, represented by negative and positive energies, collide at the speed of light, they annihilate each other and disappear into nothingness. These pairs possess an intelligence that is perceptible to humans within the realm of time. Once they transform into nothingness, we are not only unable to see or recognize them, but we also cannot comprehend their intelligence. This is because the pairs become unified and lose their distinct poles. The time domain does not accommodate unipolar existence.

Philosophically, we cannot assert that matter has ceased to exist. If it has disappeared, where has it gone? And why has it vanished? The reality is that matter has not ceased to exist; instead, we lack the tools to detect it. Within the time domain, we identify everything by its opposite. For instance, darkness is defined by light, along with numerous other dualities. Furthermore, every device we have created is based on our understanding and recognition, which limits our ability to detect nothingness.

When two charges, positive infinity and negative infinity, merge, they create the absolute power of an infinite number. However, we cannot comprehend this absolute power within the bounds of time, which we consider to be nothingness. Although the two positive charges combine into one, they become nothingness while retaining their inherent intelligence. This intelligence is the same force that brought the entire existence into being from nothingness. We understand that the two charges, A+ and A-, result in nothingness, yet zero, or nothingness, is a number that holds a unique significance for both positive infinity and negative infinity; they must arise from within it. While humans are unable to achieve this, a network of super-intelligence is capable of doing so.

We can convert two charges, A+ and A-, into nothingness. Still, we currently lack the knowledge to return these two charges from the unipolar world, or the state of non-existence, to a state of temporal dominance. While the concept of nothingness may seem empty in our thinking, it is not so in reality. Nothingness is actually a quantum vacuum, characterized as unipolar, meaning it exists outside the confines of time and space. It possesses a permanent and eternal existence and embodies a form of intelligence.

Why is it considered a type of intelligence? Because science demonstrates that from nothingness, it created existence and temporal dominance. However, we still do not fully understand this process. As intelligent beings, we humans question the nature of this intelligence and seek to understand it. This intelligence can be viewed as an infinitely intelligent superbeing operating within both the realm of nothingness and the present.

Why do we recognize this intelligence? Because we are intelligent beings. This intelligence does not come from an external source; it originates from that supreme being. Everything that exists in the universe originates from Him, and He is an integral part of it. No raw materials were used to create the universe; rather, the universe emerged from His essence, making the bipolar universe a manifestation of Himself.

Humans are bound by time and perceive their existence through its lens. Yet, they remain unaware of its true nature or the laws governing it. They often ask, "Where did the Creator come from?" This question arises from their limited understanding, shaped by the duality of their experience, as they measure everything in terms of its opposite.

In contrast, nothingness embodies intelligence and lacks material form. It is unipolar, without beginning or end, with both start and end being the same. Whatever exists—both presence and absence—exists as it is. Intelligence is not subject to the constraints of time; similarly, unipolarity transcends temporal limitations, remaining neither constrained nor governed by time. Instead, it is the essence that gives rise to existence. As for the world we inhabit, it faces destruction.

According to the first and last law, number 4, everything that begins has an end. The world is perishable, and all objects in the universe are subject to the passage of time and decay.

The universe was created from a minimal source of Tanerium mass or possibly from the infinite supermassive mass of a large atom. This creation was not an explosion; instead, the Tanerium mass began to expand from the moment of its formation under the influence of the gravity of the Paravazium layer. This expansion is still ongoing, although the rate of expansion is gradually decreasing.

The diameter of the Sun is approximately 1.4 million kilometers, while the diameter of the Earth is about 12,800 kilometers. Around 14 billion years ago, the Sun's diameter was approximately 101 millimeters, and the Earth's diameter was roughly 1 millimeter. Over 14 billion years, the Sun's diameter has increased by 101 millimeters, while the Earth's diameter has increased by 1 millimeter per year.

Initially, the Sun and Earth were made up of superatoms, which possessed a significantly higher density. There are several reasons why Earth began to solidify after its molten shell cooled, becoming thicker over time. This raises the question: Why did Earth grow so large and develop a solid shell, considering it started as a molten mass?

The Earth's molten state can be traced back to the beginning of the universe, when a mass of a hypothetical element called Tanerium was created as a subset of superatoms. These superatoms had electron layers containing trillions of individual superatoms, which later formed galaxies and stars. Originally, stars and planets were composed of superatoms and existed at infinitely high temperatures.

Superatoms generate substantial magnetic fields, enabling them to maintain stability at temperatures exceeding several hundred million degrees Celsius. As millions of years passed, the universe began to cool, causing superheavy atoms within these superatoms to split into lighter and heavier atoms. Some superheavy and superatomic atoms remained in the cores of the Earth and the Sun, where they sustained the Earth in a molten state by generating heat through their magnetic fields.

However, as the universe expanded and temperatures decreased, these heavy superatoms lost their stable state. They transformed into heavier and lighter atoms, eventually creating hydrogen, oxygen, water, and air. Thus, the Earth was formed from a complete mass fraction of the superatom Tanerium and initiated its expansion.

What is the primary factor that generates heat in planets and stars?

Superatoms primarily generate temperatures reaching millions of degrees by creating vast magnetic fields.

How did the Earth's climate and weather originate?

The Earth was initially composed of superheavy atoms, remnants of the primordial cloud that gave rise to the universe. These atoms generated enormous magnetic fields, leading to surface temperatures that reached several thousand degrees. Over time, they transformed into both heavier and lighter atoms, resulting in the formation of oxygen and hydrogen.

As water vapor formed and thick clouds developed, the Earth's crust began to cool and solidify. This process allowed rain and snow to fall on Earth for thousands of years. As the crust cooled, superheavy atoms with higher mass and density inside the Earth transformed into lighter atoms, which had greater volume and lower mass. This transformation caused the Earth's volume to increase and continues to occur.

Simultaneously, climatic phenomena developed on the Earth's surface, leading to the formation of seas and lakes. These changes created a suitable environment for living organisms, including marine animals, as well as the growth and multiplication of aquatic plants.

Over time, animals began to inhabit the land, but the Earth's crust underwent millions of changes. Earthquakes have destroyed various landscapes while new landmasses have emerged from the sea. Today, we find sedimentary mountains situated thousands of kilometers away from oceans and seas.

I was surprised to find a shell fossil while digging in a mountainous region of northwestern Iran, particularly in Azerbaijan, which is at least 1,500 kilometers from the nearest sea. The release of light gases such as oxygen, nitrogen, hydrogen, and carbon dioxide from the Earth's crust contributed to an increase in the Earth's volume. The pressure from these gases led to the formation of sedimentary mountains that rose from the depths of the ocean.

Due to nuclear changes in superheavy atoms, the Earth's crust gradually cooled, increasing the production of oxygen and hydrogen. Over millions of years, climate change has resulted in increased rainfall and snowfall, while the Earth's crust has shifted and evolved. Internal pressures from within the Earth caused the formation of mountains as the volume of the Earth's interior grew. Once the Earth's crust became thick enough, volcanic mountains developed to relieve pressure beneath the crust by allowing gas and molten materials to escape during eruptions. Even now, the Earth's volume continues to expand after billions of years.

If it is true that the Earth did not separate from the Sun about 14 billion years ago, then the diameter of the Earth was infinitely smaller—around one millimeter. Today, that diameter has grown to approximately 12,800 kilometers. While we do not know the exact size the Earth was when it formed, current scientific understanding suggests that not only is the universe expanding, but stars and planets are also expanding.

We know that the Earth's depths and core are molten, and some scientists believe that the planet's internal collapse generates this heat. However, this theory has been largely rejected and lacks scientific support. Instead, it is thought that superheavy atoms produce the heat in the Earth's core.

Superheavy atoms significantly contribute to the electromagnetic fields that generate heat, and these magnetic fields are clearly visible on and around the Earth. These superheavy atoms are remnants of the original superatoms that existed at the universe's beginning. They can only remain stable at high temperatures, making the Earth's core the only safe place for them. The survival of the Earth depends on these atoms. As cold gradually penetrates from the crust to the core, the superheavy atoms lose their stability and transform into lighter atoms, producing gases such as carbon dioxide, oxygen, hydrogen, and nitrogen. This process increases the Earth's volume and intensifies volcanic activity.

1. How do earthquakes occur?

2. How are oil and gas formed deep within the Earth?

3. How is coal formed?

4. Is the Earth's magnetic field disappearing?

When heavy atoms with higher density and smaller volume convert into lighter atoms, gases such as oxygen, hydrogen, nitrogen, and carbon form in different layers of the Earth's core; these gases seek escape routes and flow through the sublayers. As the oxygen layers move, they collide with hydrogen layers, resulting in explosions that can trigger earthquakes. Similarly, when oxygen layers combine with carbon layers, they also create explosions that cause earthquakes.

This process doesn't stop there; it generates carbon dioxide and water vapor, which continue to flow and move through the sublayers of the Earth's crust. Over time, these gases accumulate in the faults of the Earth's folded crust. Eventually, this excessive accumulation forces the gases to find new pathways, potentially leading to increased seismic activity and a higher frequency of earthquakes. Deep within the Earth, these gases continue to migrate toward volcanic mountains, from which they can eventually escape into the Earth's atmosphere.

Scientists believe that the movement of the Earth's tectonic plates causes earthquakes, a theory that has gained significant acceptance. Each year, millions of tons of gases, including oxygen, hydrogen, and carbon dioxide, are released from deep within the Earth into its atmosphere. The existence of air and water, the magnetic field, and life itself is attributed to the superheavy atoms found in the Earth's core. These atoms have existed since the universe's inception. Stars are rich in superheavy atoms, which not only generate the necessary heat for themselves but also provide energy to the planets orbiting them.

Superheavy atoms are essential for the existence of life, and their destruction on a planet has far-reaching consequences. When a planet lacks these atoms, it is unable to support air, water, oxygen, or life itself. This process unfolds in several stages. Firstly, the planet loses its active volcanic activity. Secondly, earthquakes and tremors cease to occur. Thirdly, the planet's electromagnetic field gradually diminishes, eventually disappearing over millions of years. Lastly, water and oxygen levels, along with other gases, begin to decline and ultimately disappear. Although the electromagnetic field does not vanish immediately—thanks to the presence of certain heavy atoms in the planet's core—it becomes increasingly ineffective over time due to their scarcity. Ultimately, these conditions result in a lifeless planet.

I refer to these planets as "dead planets" because they no longer have the means to support life and are in the process of destruction. Regarding oil and gas, scientists believe that these essential substances formed from the accumulation of organic materials, commonly known as fossils. However, is this truly the case, or is it just a theory?

In the Earth's central core, hydrogen, carbon, and other gases are generated. Under high heat and pressure, hydrogen and carbon can combine to form oil and gas. The belief that oil and gas solely originate from fossils has been largely dismissed. Instead, coal is considered to form from carbon veins resulting from nuclear changes in superheavy atoms. If these carbon veins are subjected to heat and pressure deep within the Earth, along with hydrogen, they could potentially transform into oil.

Natural gas is used as fuel, whether in liquid or gaseous form, and is often located near oil reserves. Superheavy atoms are thought to be essential for the existence of life on Earth. Furthermore, there is no scientific evidence to support the idea that all fossils are buried deep within the Earth and have been transformed into oil and gas through this process.

Why do stars explode?

The element hydrogen plays a significant role in the life cycle of stars. In previous discussions, we explored the law of nuclear change, and now we will delve a bit deeper into it.

At the beginning of the universe, it existed as an infinitely dense mass. However, its exact volume and diameter remain unknown. This mass can be thought of as a "super atom," a collection of trillions of super atoms arranged in the electron orbits of seven orbital layers. As the universe expanded, this super atom began to break apart, leading to the emergence of super atoms, each of which eventually formed galaxies. These super atoms were composed of even heavier and superheavy atoms at their cores.

The universe continued to expand, ultimately giving rise to the formation of stars. These stars are composed of super atoms, which represent multiple generations derived from the cores of these extremely heavy atoms.

The superatoms began breaking apart, resulting in the formation of heavier atoms, followed by lighter ones. This process ultimately gave rise to elements, with hydrogen being the most significant as it is the lightest element in nature.

Hydrogen accumulated in the outer layers of stars, where it burned under the influence of strong electromagnetic fields, transforming into helium. This reaction generated the heat necessary for the stability of superheavy atoms on the surfaces of stars.

As hydrogen was converted, other heavy atoms, including iron and various metals, originated from the superatoms and settled into the lower layers of the stars. Over millions of years, the quantity of heavy atoms significantly increased, forming thick protective layers around the superatoms. Consequently, the stars turned into infrared sources; although their outer layers were hot and molten, there was no hydrogen fuel available. This was because the superatoms could not escape the dense protective layers, which prevented them from producing hydrogen for fuel.

While hydrogen production continued beneath the protective shell of the star, it could not escape. Over many years, vast layers of hydrogen accumulated as a mass or liquid in the lower parts of the shell, becoming pressurized until they ultimately exploded in the presence of a strong electromagnetic field. This resulted in a hydrogen bomb explosion occurring beneath the star's outer shell of infrared radiation.

In this powerful explosion, a portion of the star broke away and formed a molten sphere, dominated by the star's gravity, which caused it to orbit the star. The size and mass of the newly formed planet depend on where the explosion occurred within the star's shell. If the explosion occurred deeper in the shell, the planet would be more extensive; whereas if it occurred shallower, the new planet would be more minor.

The star does not perish ultimately but continues to exist, radiating heat and light while utilizing its remaining hydrogen fuel, although its initial mass will be reduced.

The planet contains a substantial amount of superheavy atoms, which typically accumulate at or near its center. This planet has an electromagnetic field and continues to exist, gradually cooling as it does. It is also believed to have elements such as water and air, as it can generate hydrogen and oxygen using its superatoms. Whether this planet can sustain life depends on its location and size.

The new star it orbits will persist for millions of years but may eventually experience further explosions. Our sun has exploded at least ten times, each time ejecting a planet. The most recent planet to separate from the sun was Mercury. With each explosion, the star becomes smaller, potentially leading to the formation of more minor planets in subsequent events.

Ice Ages on Earth

Earth, like other planets in the solar system, was formed after a hydrogen explosion separated it from the Sun. Throughout its history, Earth has experienced two significant ice ages. After the separation, the planets enjoyed the warmth and light of the Sun for millions of years. However, heavy atoms produced by superatoms on the surface of the Sun formed a thick shell that prevented hydrogen from reaching the Sun's surface. During this period, the Sun entered an infrared state, which likely lasted between one hundred and one hundred fifty thousand years, resulting in a severe ice age across the solar system.

As hydrogen production intensified deep within the Sun's core, a massive explosion occurred due to the resulting pressure and heat, coupled with the electromagnetic field created by the superatoms. This explosion resulted in the creation of Venus. Following this event, the Sun emerged from its infrared state, once again providing light and energy to the surrounding planets. However, this cycle was not an isolated event; the Sun underwent this process again, leading to the formation of Mercury.

Throughout its history, Earth has faced the Sun's infrared state twice, experiencing extreme winters each time. This cycle is expected to continue, with the Sun eventually transforming into a superatomic star, retaining only its core, which would measure about 20 kilometers in diameter. Physicists mistakenly refer to these as neutron stars, but they are better described as superatomic stars. These stars eventually explode after millions of years, transitioning into black holes or, in some cases, becoming comets. The smallest element in nature, hydrogen, is a significant force capable of causing widespread destruction throughout galaxies and massive stars. It is crucial to recognize that hydrogen has a destructive impact on the universe.

The conclusion of Earth's last ice age could coincide with the birth of Mercury. Another pivotal event for the Earth was caused by a hydrogen explosion deep in the Pacific Ocean, which ejected the Moon and led to the formation of the Tibetan Plateau and the Himalayan Mountain range. This explosion also shifted the Earth's poles, resulting in a substantial increase in temperature and the near extinction of most living organisms. This geological activity occurred primarily in the Pacific Ocean and Southeast Asia, where the Earth's core is slightly closer to the crust than in other regions, leading to many dangerous earthquakes.

The transformation of "superstorms" with high density and low volume into "lighter atoms" with greater volume has contributed to the expansion of the Earth. Consequently, Earth has undergone numerous changes over millions of years to achieve its current state and beauty.

What are black holes? What are the Comets?

The final stage of a star's life occurs when it explodes and no longer exists as a planet. This occurs when the star's mass decreases, leading to an explosion in the core that causes the star to collapse and transform into a hollow sphere. The center of the star becomes empty, and the materials from the explosion rotate in a ring or hollow sphere formation. Eventually, these materials drift apart and can become asteroids or, in some cases, comets. Comets form when multiple sequential explosions occur near the core, triggering further explosions that align with each other. In this scenario, the resulting structure does not turn into a black hole but instead becomes a comet that travels through space.

Some supermassive black holes form from the explosions of central stars in galaxies and are significantly larger than typical black holes. These black holes primarily consist of dark matter and are located at the centers of galaxies, where massive layers of hydrogen can ignite explosions resembling hydrogen bombs in their cores. This phenomenon remains largely unknown to physicists.

The core's central region contains superheavy atoms that, after the explosion, are distributed throughout the hollow sphere and exhibit a much greater gravitational force. Following the explosion, the material they eject does not get pulled into the center of the black hole, as that area is completely empty. Instead, the gravitational force acts only on the periphery, allowing the expelled pieces to be attracted to the surrounding regions of the black hole.The Paravazium Layer , known as hidden energy,acts like a super black hole surrounding the universe,as any light that hits it cannot escape .

Why do some planets explode?

Some larger planets, or those that contain significant amounts of superheavy atoms, can accumulate hydrogen gas inside them after their outer layers cool, especially when they separate from their main star (as is the case with Earth). This buildup of hydrogen gas can create conditions that lead to an explosion. I believe that the Moon was formed as a result of such an explosion from the Earth. If the Moon had separated from the Sun, it would have become a small planet orbiting the Sun instead of being a satellite of the Earth. It's interesting to note that the Moon moves away from the Earth at a rate of about 4 centimeters per year.

Why did the planet Damavand commit suicide in the solar system?

The solar system consists of nine planets orbiting the Sun and is located in the Milky Way galaxy. Their distances from the Sun are: Mercury—57 million kilometers, Venus—108 million kilometers, Earth—150 million kilometers, Mars—228 million kilometers, Jupiter—779 million kilometers, Saturn—1.43 billion kilometers, Uranus—2.88 billion kilometers, Neptune—4.5 billion kilometers, and Pluto—5.91 billion kilometers.

The distances of Venus, Earth, and Mars from the Sun each increase by roughly the same amount as the distance of Mercury. However, Jupiter's distance from the Sun is over three times that of Mars, raising some questions. Based on the distances of the first four planets, Jupiter should be approximately 300 million kilometers from the Sun, not the actual 779 million kilometers. Furthermore, according to the distances of the next four planets, it should be around 450 million kilometers, which still presents a significant discrepancy.

This suggests the intriguing possibility that a planet may have existed between Mars and Jupiter, which was destroyed, possibly by a hydrogen explosion. The remnants of this planet are currently scattered throughout the solar system as moons and planetary debris. Currently, Jupiter has 67 moons, Saturn has 62, Uranus has 27, Neptune has 14, Mars has two, and Earth has one, totaling approximately 170 moons, a number that is likely to continue growing. Additionally, there are countless asteroids within the solar system, and thousands of them fall to Earth each year, although their exact quantity remains unknown.

The existence of all the moons and rings surrounding Jupiter, alongside the rocky bodies of the planets and a noticeable gap between Mars and Jupiter, suggests that there was likely a planet in that region. This planet seems to have met its destruction. Furthermore, the presence of Io, a moon with a diameter of 3,643 km and the only known volcanic moon in the solar system, implies that the core of this planet has been significantly impacted. I have named this hypothetical planet Damavand, after the highest semi-active volcanic peak in Iran, which is located east of Tehran.

It is improbable that a moon of such a small diameter possessed superheavy atoms. However, it remains plausible that this moon could have supported a climate. A planet or moon with superheavy atoms can generate a strong magnetic field and sustain climatic conditions. Therefore, it is reasonable to consider Io as the remnant core of Damavand, which may have once been one of the largest planets in the solar system, orbiting the Sun between Mars and Jupiter.

Based on my calculations, the distance of this planet from the Sun is approximately 350 million kilometers. Given its characteristics, it is not implausible that this moon could also harbor life.

A Discussion in the World of Metaphysics

For thousands of years, humans have believed in the existence of the soul, while others have rejected this idea. Those who deny the existence of the soul often lack sound scientific reasoning, and I find little common ground with them. My focus is on discussing this topic from the perspective of quantum physics. However, it's worth noting that those who do believe in the existence of the soul may also lack scientific reasoning.

In this section, I aim to explore the existence of the soul. Drawing on my 60 years of experience in philosophy and my interest in quantum physics, I will strive to present my argument within a scientific framework. Whether or not prominent professors and scientists accept this argument is secondary to me; what matters is adherence to scientific standards.

I have come to realize that human awareness exists along a continuum, ranging from zero to infinity, across various fields of inquiry. Education, along with religious and scientific prejudices, plays a significant role in shaping this awareness. However, the main factors influencing human awareness and the lack thereof involve deeper issues, which I will discuss in the following section.

First, we need to understand the fundamental structure and nature of the soul, as well as its origin and development.

Definition of the Soul

The soul can be understood as a form of intelligence. Two types of intelligence reflect the bipolar nature of the world. The first type is physical intelligence, which is present in all particles and objects in the universe. Each particle, characterized by its two poles—negative and positive—exhibits feelings and has needs. Negative charges are attracted to positive charges, while positive charges are attracted to negative charges to fulfill their needs. This inherent need within matter suggests a form of feeling, which I refer to as intelligence. Therefore, feeling and intelligence are fundamentally the same concept. Every particle in the universe possesses both feeling and intelligence, and this will always be the case.

Unlike inanimate objects and plants, living beings, such as animals, possess two types of intelligence: physical intelligence and metaphysical intelligence.

Now, we will discuss the difference between physical and metaphysical intelligence.

Physical intelligence refers to the inherent qualities embedded in the material structure of living beings. This can be understood as the genetic makeup of an individual, which defines various characteristics such as height, body shape, dietary needs, suitable living environments, reproductive methods, and more. While a living being possesses all the attributes of life through physical intelligence, it cannot think, plan, or feel, which are aspects of metaphysical intelligence.

For example, when a person is asleep, their body utilizes its energy reserves to ensure that vital organs, such as the heart and digestive system, function properly. Although the individual is alive, they are not conscious or aware and, therefore, does not exhibit metaphysical intelligence. It is only when a person becomes conscious that they access their metaphysical intelligence.

When a living being or human is asleep or unconscious, it is often considered that they do not possess a soul or metaphysical intelligence. Consciousness in a living being is referred to as the soul. This metaphysical intelligence is typically accompanied by physical intelligence when the body is in perfect health. However, if there is a defect in the physical body, the metaphysical intelligence—or soul—may withdraw.

This raises the question: Why does the soul seem to separate from the body during sleep? Sleep can be viewed as a temporary state that affects the brain and nervous system, resulting in a temporary halt in activity and awareness. Although this state resembles a kind of disease, it requires only rest, during which the seven senses shut down. After a few minutes or hours, consciousness returns, revealing the individual's intelligence or soul.

The universe itself exhibits physical intelligence, which governs the movements of planets, stars, galaxies, and other celestial bodies. The existence of plants and animals can also be attributed to physical intelligence; however, without metaphysical intelligence, they lack reason and will.

The primary distinction between living beings and inanimate objects is that living beings possess consciousness, reasoning, and will—often referred to as the soul or metaphysical intelligence. In contrast, inanimate objects lack these qualities. Physical intelligence pertains to the bipolar world governed by time, whereas in the realm of nothingness, physical intelligence does not exist. Nothingness is where metaphysical intelligence resides, serving as the creator or source of time and existence. The only true superintelligence exists in the metaphysical realm, which can design, plan, and create existence—capable of bringing it into being, destroying it, and recreating it.

Design and planning are fundamental to a metaphysical intelligence network that permeates all existence, containing infinite information and the ability to shape the genes of countless organisms and plants. In contrast, physical intelligence concentrates solely on its own needs and is limited by the will of metaphysical intelligence. This metaphysical network exists within everything, including even nothingness, yet we lack the means to perceive it due to its unipolar nature.

In the material realm, we do not encounter a tangible unipolar mass or particle, although such entities exist beyond our recognition. The indistinguishable mass or matter represents the very essence of nothingness or the metaphysical intelligence network. While nothingness can assert its existence in the physical world, we remain unable to perceive it.

Unipolar refers to the absence of an opposite pole, leading to a lack of discernment. We remain unaware that existence is enveloped in nothingness or that nothingness has overtaken existence. We struggle to recognize both nothingness and unipolarity. I have presented the necessary information, supported by scientific reasoning. Those who do not grasp these concepts should revisit the book multiple times for a complete understanding.

Who is God? What is His nature? And where did He come from?

The Greek philosopher Aristotle, who lived in the 4th century BC, shared some profound insights about God. He stated, "God has created laws in the universe, and we humans measure existence using the same laws that God established. If we separate ourselves from this universe of time and space, we will see that there are things in which an infinite force has been replaced, capable of creating existence." Although Aristotle made these observations around 2400 years ago, they remain relevant today.

When Aristotle refers to "laws," he discusses the fundamental nature of time, a concept I have noted repeatedly throughout this book. We are governed by time, arising from paired charges that help us understand existence by contrasting one pole with its opposite. For example, darkness and light, along with attraction and repulsion, are opposing forces that define each other. Darkness enhances our comprehension of light, and attraction is understood through the presence of repulsive forces.

The challenge is that humanity still does not fully grasp the nature of time. We often fail to recognize that the constraints and dynamics of time itself influence our questions and answers.

Humans still do not understand that nothingness is a transcendent concept that exists outside the realm of time. When I refer to nothingness as a "concept," I mean that it lacks both temporal and spatial dimensions. When we discuss nothingness, we are indeed talking about nothingness itself. This is not an absolute emptiness, but rather a quantum void. From this perspective, we cannot determine the origin of nothingness, and asking such a question is meaningless.

On the other hand, we know that the universe emerged from nothingness and will eventually return to it. According to the twenty laws governing the structure of the universe, everything that has a beginning must also have an end, suggesting that the world is in a constant state of flux, with everything in a perpetual state of disappearance.

After more than 60 years of scientific research, I have concluded that the essence of time is rooted in the changes of energy. For the first time since humanity emerged on Earth, I am presenting this idea to the scientific community. All seekers of knowledge and understanding must be aware of this definition. While time ultimately destroys all existence, it can also be eradicated through the destruction of matter, leaving nothingness or non-existence to remain.

According to the 20 laws governing the structure of the universe, the entire cosmos is composed of an infinite number of positive charges and an infinite number of negative charges. When these two forces are combined, they result in a state of nothingness, represented by the number zero.

However, philosophy offers a different perspective. It suggests that the concept of zero, or nothingness, is not and will never be absolute nothingness; instead, it exists beyond the bounds of time and space.

We have two clear reasons for understanding the nature of the universe's creator. The first reason is intuitive: the operator of our existence and the creator of all living beings and plants is an infinitely intelligent being who designed and introduced billions of genes that contribute to the diversity of life. As humans, we are among their witnesses and observers. If you examine your own body, you can observe the complexities and scientific principles that underlie its structure and function.

I have previously discussed the intelligence found in objects and particles, which, although they cannot design or structure living plants and animals intelligently, possess the ability to sense and understand their needs. Some people believe that the entire order of existence, including plants and animals, emerged from nature itself. However, they do not recognize that nature and existence are manifestations of God and the Creator, who fashioned all of existence from His essence. There were no raw materials outside of Himself that could be used to create the universe.

We have previously discussed this topic, and the 20 laws regarding the structure of the universe provide essential explanations for it.

We are intelligent beings living in a world filled with intelligence, yet we often remain unaware of our intelligence. It's similar to a fish in the depths of the ocean that doesn't realize water exists because it has never seen beyond its liquid environment. Our human genes were shaped by a superintelligent universe, designed to create living beings with specific physical characteristics. This intelligence permeates all of existence.

Our physical bodies lack their own will, intelligence, and consciousness; they require metaphysical intelligence to become alive and functioning entities.

Without this intelligence, we would be unconscious and motionless. For instance, a person in a state of sleep lacks awareness and consciousness. However, when intelligence, or a "soul," awakens within them, they can become wise, intelligent, and capable of making choices.

The soul, or metaphysical intelligence, in living beings is timeless and eternal. Humans do not die; instead, their bodies cease to function. For millennia, people have misunderstood this fact, equating death with the body's failure, as society tends to identify individuals by their bodies rather than their souls. The body serves merely as a tool for the soul, which is a creative and everlasting intelligence. It may take time for humanity to fully grasp this understanding.

Each person possesses a unique soul, perfectly aligned with the cells of their brain and body, which monitor the body throughout life. Souls do not inhabit one another. While people believe their bodies experience joy, it is actually the soul that feels it. Similarly, the soul is responsible for thinking and decision-making, using the brain as an instrument.

When the body tires or suffers, the soul recognizes the need for rest and temporarily detaches, remaining connected until the body heals. Humans are, in essence, their souls; the body is merely transient. After numerous physical transitions, the soul achieves perfection and does not return to Earth; instead, it assumes a new body in another realm, continuing its journey. People can live there for millions of years, free of lies, deceit, and conflict, and all beings can coexist in harmony.

The universe can vanish without a trace, a cycle that occurs infinitely across countless universes, each one emerging and disappearing in succession.

This leads to the inquiry about the nature of the creator or creators behind it all.

The Creator exists in nothingness, which is neither material nor temporal, as He creates time Himself. The Creator embodies intelligence, which is immaterial in nature. At the end of the book, we will provide a scientific perspective on the nature of God. Nothingness is not empty but a quantum vacuum composed of matter that transcends time and space. All existence and nothingness are encompassed by intelligence. When we refer to nothingness, we mean the absence of material, or quantum nothingness; this nothingness contains cells that are not material.

There is space, but it consists of a quantum vacuum. The term "quantum" refers to particles. When the Paravazium force cloud layer retreats, it does not push back against nothingness; instead, nothingness permeates the entirety of existence. In fact, all existence is immersed in this quantum vacuum. Space, which is characterized by nothingness, or the quantum vacuum, is fundamentally different from space, which is dominated by time. The quantum vacuum is devoid of matter.

The physical aspect may be destroyed, but according to the concept of reincarnation, the metaphysical aspect is eternal and will be reborn. Intelligence is akin to a driver who continually changes vehicles. Every human possesses a soul of intelligence, and this intelligence never dies.

What is the purpose of human existence?

Why are some people good while others are bad? Why can't some individuals find the truth and become guided? What is the essence of humanity? What does self-knowledge involve? What role does God play in human existence? Was humanity created through free will or compulsion?

In this section, we will explore the philosophical dimension. I am not affiliated with any religion or sect. After years of research on this subject, I have uncovered some truths that I wish to share with humanity.

It is essential to focus on the present. For God, or the Creator of the intelligent network of the universe, the concepts of past and future hold no significance. This intelligent network operates without the need for thought, review, or modification; it consistently executes the most intricate and significant affairs of existence in the present moment.

When humans immerse themselves in the present—free from the material concerns of the past and future—they connect with the intelligent network of the universe. In this state, individuals can begin to comprehend the truths of existence, limited only by their capacity for awareness and understanding. This enlightened state represents a form of consciousness. Awareness embraces those who dwell in the present moment, allowing the intelligent network to share its knowledge and insights with those who draw near to it.

Closeness to God is shown through actions, not just words. You can connect with the intelligent network by embracing the entire universe and demonstrating kindness to all living beings. Kindness and compassion are among the highest qualities appreciated by this network.

Consider two individuals: one may not believe in God at all but lives in alignment with divine principles by being kind and doing good for others to the best of their ability. In contrast, another may claim to have faith in God but consistently engage in wrongdoing and mischief.

You would likely regard the first person more highly than the second. The intelligent network of the universe also favors the first individual for their acts of kindness, while it stands against the second for being deceitful and misguided.

Every living being, whether animal or human, possesses a soul or intelligence. It is essential to understand that the soul or intelligence first enters the body. While our soul or intelligence is present in humans, it remains invisible to us. The eternal soul lacks sufficient experience and does not coexist with a new body over time. The purpose of the soul entering the body is to facilitate the evolution and advancement of the soul itself, not the body. The body is material and cannot evolve or advance.

As the body grows, it encounters many demands. An animal's soul, after wandering and living for a long time, accumulates experiences until the life of the body comes to an end in death. After this, the soul or intelligence transitions to a different body, taking its previous experiences with it. The soul continues to change bodies many times until it eventually reaches a human form. The soul that enters a human body is fundamentally different.

Animals do not possess free will, but humans do, along with a profound soul that has accumulated many experiences. A crucial point to understand is that people often confuse themselves with their bodies. When faced with trouble, difficulties, or suffering, they believe their bodies are the ones in pain.

However, this is a misconception! What indeed suffers is your soul, not your body. The body itself lacks feelings; it is the intelligence or soul that provides feelings to the body .

When you are asleep or unconscious, your intelligence or soul is absent, which is why you do not feel anything during those times. This indicates that sensations of pleasure and feelings of good or bad originate in your soul, not in your body. Essentially, you are the soul or intelligence, not merely the body. Your body serves as a material tool that is under the control of your soul. In essence, the body is at the service of the soul as a compound entity.

The body has a tangible and physical structure that is visible, while the soul has a metaphysical nature that is not. Although the soul possesses intelligence, it does not share the same level of intelligence as the Creator, as it has not experienced the passage of time and has not reached the highest level of advancement. Souls exist in a state of nothingness, characterized by absolute silence and darkness. Their greatest desire is to gain control over time and experience worldly pleasures.

You are not merely your body; you are the soul that has willingly entered the world, driven by countless desires for enjoyment. You should not complain about this existence. Any protests arise from the hardships of life that you have created for yourself through your own actions. These difficulties stem from unjust and unwise desires formed in past lives. God provides guidance but plays no other role in your life.

Before achieving perfection, the soul seeks to enjoy life and often encounters difficulties through its misguided pursuits. These experiences allow the soul to learn and move toward perfection by finding the right path and gaining awareness.

A person's life is like a boat being carried by a river toward the ocean. Inside the ship, he has complete freedom to choose between doing good or bad things, but what he may not realize is that the captain of the boat represents his actions. If he decides to do good, he will eventually reach the shore of peace, and gradually, he will gain more control over the boat, allowing him to guide it toward beautiful destinations by his own will. As a result, life will feel free for him.

However, if a person engages in evil actions, like putting holes in the boat and fighting against the captain, he risks being drowned in the ocean. In this scenario, he must accept whatever challenges come his way.

Education plays a vital role in human life, and every individual is expected to contribute according to their abilities, strengths, and capabilities. This is the stage where a person begins to shape their destiny. If he is inclined to do good, he will accumulate positive actions for himself; if he tends toward evil actions, he will accumulate negative consequences instead.

It takes a long time for the body to fail and for a person to reach the stage of death ultimately. At that point, the soul enters another body, carrying with it the weight of its actions. A person's past actions define their essence and are considered their destiny.

If someone has performed good deeds, those actions will encourage them to continue doing good. If they are conscious of this, they will consistently strive for goodness. Conversely, if they commit mistakes and engage in wrongdoing, they will eventually have to face the consequences.

A person must accept and endure unfortunate events that cannot be solved in order to move through them. This journey is essential for elevating the soul. Each time a person is reborn, their awareness grows until they reach a stage where they no longer make mistakes.

At the end of this process, the soul separates from the body and enters another realm where people coexist in complete peace and love.

Human life is very brief, and everything is ultimately in the hands of individuals, who determine their own destinies. The entry of souls into the bodies of living beings, including humans, is voluntary and not coerced. Some people accumulate more bad deeds than good, and this imbalance is carried with their souls. When they enter a new body, their past actions may provoke them to repeat harmful behavior, often without sufficient awareness of their choices.

No one else is to blame for this; each individual is responsible for their actions, as they have the freedom to choose between good and bad paths. Unfortunately, some people pursue their own desires at the expense of others, which undoubtedly carries consequences.

In this life, or in subsequent periods, such individuals may find themselves trapped in a cycle of trouble and struggle, often unaware that they are paying the price for their past actions. Some realize they are dealing with the consequences of their previous choices, while others express disdain for existence and blame God and others for their suffering. They fail to recognize that God has granted them the authority to make their own decisions and is not responsible for their actions.

Additionally, some wrongdoers engage in cruelty, injustice, and crime throughout their lives until their end. In such cases, their situation changes dramatically: they can no longer return to life as humans. Instead, their authority is stripped away, and they are placed into nature as animals, often persecuted and preyed upon by stronger creatures, forced to confront the consequences of their actions.

Many animals recognize that they were once humans, but they lack the language to express this. Intelligent and compassionate animals are among those aware of their past. It is essential to emphasize that no person has the right to mistreat animals, as such actions can lead to a reversal of roles, where the oppressor becomes the oppressed. Everyone is responsible for their actions. We must be kind and compassionate toward all beings and assist animals whenever possible.

God does not oppress anyone; rather, it is humans who create their own suffering. The Creator has established an order within the system of existence, providing humans with reason and guidance. However, this guidance is available to those who seek it. Willingness translates into action, not mere words. God is not responsible for human actions, nor for their punishment or mercy. When I refer to humans, I am speaking specifically about the human soul, not the physical body.

Many people blame God for misfortunes and negative experiences in nature, but this perspective is misguided. Ultimately, each individual is responsible for their own actions. God is not accountable for human behavior. For those who can comprehend it, the benefits of consciousness are ever-present in the world. The reason some individuals possess consciousness while others do not relate to the nature of humanity, which depends on one's past actions—achievements that come from personal effort rather than divine intervention.

Awareness does not depend on education, qualifications, age, wealth, status, or money. I observe that many educated individuals remain ignorant, and a significant number of politicians are entirely misguided, leading to the suffering of millions for the sake of money, status, and selfishness. Such harmful actions drive them toward evil and ultimately lead to their own destruction, which may result in their devolution into more base forms of existence, suffering for aeons.

We encounter individuals in society that can be described as "scientific animals," "political animals," and "capitalist animals"—people who are merely inhabiting human bodies during this lifetime while their true nature remains animalistic. Only time will reveal their fate in future existences. It is important to note that no one, not even God, determines the path of humanity; it is up to each individual to clarify their own purpose.

Some people occasionally become aware of their past lives. For instance, they might recognize individuals they've never met or visit places that feel familiar, even if they've never been there. This phenomenon is connected to their previous lives. Humans carry not only the memories of their past actions but also their innate talents of earlier incarnations.

A person who was a doctor in a past life may develop a strong interest in medicine in this life, eventually becoming the best doctor in the world with dedication and education. Similarly, someone might excel as an economist, musician, painter, mathematician, physicist, or philosopher based on their prior experiences.

It is essential to allow children to pursue their own interests rather than forcing them into specific fields; everyone possesses unique talents and intellectual abilities. Often, a soul will reincarnate in a familiar location, although sometimes it may move to a different country. It's also possible for roles to reverse within families; for example, a father in one life may become the child in the next, or a person who lived as a woman may return as a man.

Consciousness is shaped by human behaviors, which are part of our nature. However, we must recognize the significant influence of education on this process. When Jesus Christ spoke of his coming, he referred to a transformation of the body, having reached a level of consciousness that allowed him to perceive the existing order in the world. He did not express this explicitly, as the people of that time lacked the capacity to understand.

The increased intelligence of new generations can be attributed to the fact that they have undergone many bodily incarnations and possess heightened talents. These newer generations tend to be more truthful, avoiding lies, hypocrisy, arrogance, self-importance, and excessive accumulation of wealth. Our planet is evolving towards greater intelligence and sincerity.

Future generations will have access to new sciences and will likely experience more peaceful lives. They will view the wars, bloodshed, and killings of today as the result of contemporary humans' ignorance and folly, perhaps even deeming us savages. By the end of the 21st century, many individuals will travel from Earth to other planets. Additionally, the study of metaphysics on Earth is rapidly advancing.

Heavenly Books

We humans often believe that we are the only intelligent beings inhabiting the universe, but this is not the case. In the supernatural world, there are many souls whose knowledge far surpasses our own. These beings frequently travel to and from Earth, and with their advanced understanding, they can render themselves invisible, remaining hidden from our view.

Among them are the righteous individuals who practice monotheism and have brought heavenly books to Earth. Their purpose is to guide us away from idolatry, sun worship, and star worship and encourage us to embrace monotheism instead. They do not interfere in our lives, yet they possess the power of induction and have significantly contributed to the advancement of science.

Human genes were likely manipulated to adapt to the planet's climate and then transferred here. However, the sacred texts were altered by individuals in medieval times who had vested interests in slavery, murder, plunder, and rape. As a result, these texts seem to provide a false justification for violence and exploitation in the name of God. This is entirely contrary to the actual attributes of God, which are love, compassion, and kindness.

Those who accuse God of such characteristics portray the Creator as a savage and bloodthirsty being, even while they simultaneously speak of God's love and affection. The existence of these contradictions indicates that parts of these texts have been manipulated. Some individuals exploit sacred scriptures for personal gain, twisting God's message to further their agendas. They ignore God's love and focus instead on verses that promote anger, wrath, and violence, falsely attributing these messages to God in order to justify their crimes while considering themselves His representatives.

Such people do not honestly know or accept God. They use religion as a mere tool for their profit. By promoting these false views of God, they become adversaries of the divine, seeking to undermine belief in God entirely. Their motivations are self-serving, and they disregard the core teachings that call for goodness and condemn evil. God emphasizes personal responsibility, stating, "I have given people freedom, shown them the right path, and it is up to them to choose."

In sacred texts, a liar is deemed an enemy of God. The verses within these texts are intricately woven, much like the threads of fabric. If any falsehood is introduced, it becomes self-evident and exposed. Of course, not everyone can recognize these falsehoods without divine guidance. Throughout history, the righteous are often referred to as angels.

The other group consists of those who reject monotheism and God altogether. They have come to Earth and engaged with fallible individuals to instigate oppression and violence, presenting themselves as prophets or angels while promoting crime and tyranny.

We refer to them as elves and devils, but in truth, they are merely humans from another realm. We should not be deceived by them. Their knowledge surpasses what we currently understand, though we may reach their level in time. These individuals possess powers such as teleportation and the ability to transform into any form and speak any language.

Evildoers often lose their power when faced with high-ranking individuals. They lead those who are vulnerable and mentally prepared for harmful choices down a path of deviation through encouragement, temptation, and subtle persuasion. In rare instances, they may present themselves as benefactors, claiming to be angels or prophets, to gain control over their targeted individuals. If successful, the evildoer's spirit can take over, leaving the person's original spirit as merely an observer without the ability to act.

In this state, the possessed individual commits harmful deeds that are illogical and inhumane. The most effective solution is to consult a psychiatrist who can prescribe medication to alter the condition of the brain cells. However, the challenge arises because the person, under the influence of the evildoer's spirit, often refuses to seek help or treatment.

This situation can also occur in individuals who have experienced prolonged depression, leading to mental disorders. It is essential for anyone facing such issues to consult a doctor, even if it requires persuasion. The evil spirit that has taken hold is often very intelligent and cunning. Typically, these problems stem from feelings of loneliness, despair, and excessive sadness.

While I am neither a doctor nor a biologist, it seems to me that deficiencies in specific nutrients and elements in the brain can contribute to mental disorders, making consulting a psychiatrist a top priority.

The leading cause of the decline in belief in the One God is the mixing of religions with superstitions, along with a lack of a correct understanding of divine commands. These commands have undergone constant manipulation and changes over time. Although the preachers of religious traditions can distinguish between good and evil, many have not considered or will not consider reforming their beliefs. This reluctance stems from the fact that doing so would disrupt the groups they have formed around themselves, which contributes to their material well-being. Others lack the courage to pursue reform.

Humans have an inherent tendency toward introspection and esotericism, making them easily influenced by superstitions. In fact, many people may find superstitions more relatable and appealing than the truth. This situation arises from human ignorance, which is often rooted in their nature and derived from past actions. It is important to note that humans cannot be entirely blamed for their lack of knowledge. The solution lies in providing education and implementing societal laws to curb harmful behaviors. Over time, this approach can help individuals gain experience and attain the necessary knowledge.

Philosophy teaches that the best way to undermine the value of a person or a belief system is to attribute lies that closely resemble the truth of that system. Extremists within the religions themselves often perpetuate this distortion.

The souls that leave the physical body are fully aware of all their actions and cannot protest; they must accept the consequences of those actions, regardless of the form or fate they assume in the cycle of life. After death, a person takes with them only two things: their good actions and their bad actions. When they return to the cycle of life, they bring both, and they must atone for their bad actions so that their souls can grow brighter and more refined, moving toward perfection and elevation.

There is no afterlife where God has a crematorium. People face the consequences of their bad actions in this world because their mistakes occur here. Anyone who denies God's justice is mistaken. All human actions are recorded, and the universe maintains an interconnected network of intelligence that is far beyond human comprehension. The universe operates under order and law, which must be followed precisely—no more and no less.

I am pleased to share the results of my research, the culmination of 60 years of dedicated effort, as a duty to humanity. Through the lens of quantum physics, I have uncovered insights that I hope can benefit you. This information was not adapted from existing literature; rather, it was a gift from the super intelligent network, and it is my responsibility to publish it. At this crucial point in human history, there is a pressing need for knowledge.

I understand that some may oppose this information, not out of animosity toward me, but against the concept of the Creator and the clarification of the nature of the universe's Creator. This book presents a unique, scientific approach to proving the existence of God—a first of its kind throughout human history. However, it will hold little value for those who oppose God, as they are focused solely on their material existence .

Many do not want humanity to reach a higher level of awareness, for ignorance is the only means by which they can exploit others. Each reader will derive insights from this book in accordance with their inherent awareness; some will grasp it better than others, while some may not be permitted to read it at all due to their past behavior. Poor conduct can hinder awareness, leading individuals to create trouble and problems for themselves through mistakes and errors.

It is through enduring these challenges that the soul recognizes its missteps and is guided toward the right path. Every mistake can serve as a gateway to hardship and greater awareness.

Dear readers! You might encounter some ideas in this book that you disagree with at first, but I encourage you to reflect on them over the days and months to come. You may come to recognize the truth in them. Knowledge is not bestowed upon us all at once; instead, it is given gradually, depending on our eagerness and efforts to seek it.

No matter your position or the degrees you hold, it's essential to remain humble and open-minded, for you need this information and knowledge. When you engage with the world positively, knowledge will find its way to you, often without the need for a teacher or guide. It might appear as a headline in a newspaper, a comment from someone on a bus, a fleeting idea heard on the radio, or even a tip from a stranger passing by.

Knowledge can come to you in many forms. The entire universe is woven into a vast super network of intelligence, and we often overlook it or fail to tap into it. Every moment is a chance for insight, guiding you in unexpected ways. Stay open to learning; the universe is always teaching us.

You should understand that the intelligent network of the universe loves you even more than you love yourself. It is aware of you, even if you have been oblivious to it. Engage with it and communicate with it—there is no one closer to you than this network; not even you can know yourself as well as it does. This infinite super network of intelligence has created humanity and the universe with love and care. Treat all beings with kindness, including humans, animals, plants, and nature. Remember, the time of reckoning will come swiftly. When it does, you won't be able to claim that you were ordered, excused, or forced to act in a certain way.

The laws of the universe are strictly enforced, so there should be no doubt about the consequences of one's actions within the system of existence. The intelligent network of the universe—whether referred to as the Creator, the Originator, or God—has established these laws, which operate like a double-edged sword.

On one edge is love and compassion for all beings, rewarding good deeds and virtuous actions. The other edge represents anger, wrath, and retribution against oppressors and wrongdoers. This set of laws is infallible; it not only guarantees rewards for good but also shows no mercy towards those who commit injustices. Even God does not intervene in this system, and there is no favoritism in its application.

When humans change bodies, their actions are carried over to the new body because the new body is not structurally the same as the soul that inhabits it; in other words, it is not a set-up. This means that memories from the previous life cannot be recalled and should not be remembered. However, there are instances when a person may feel a sense of familiarity, recognizing people or places they have never encountered before.

No individual can interfere with the laws of existence except for the Creator, or God. The Creator does not intervene in the system that has been established in existence. Quantum physics provides clarity on the consequences of actions. Although the soul is a metaphysical phenomenon, once it enters the physical world, it must adhere to the laws of time control. According to the principles of physics, every action it has taken must result in an effect: good for good and punishment for bad. This law is fundamental to existence, and there is no way to escape it.

What are intellect and wisdom?

When we discuss intellect, we refer to intelligence. In Persian, wisdom translates to knowledge and awareness. Here, we are examining intellect in the context of science and quantum physics, exploring its role within these fields. I have previously provided essential insights on this topic, and I will reiterate: According to the 20 laws governing the structure of the universe and the third law of the bipolar world, the essence of a living being comprises two poles—material and immaterial. This means that objects and particles consist of both positive and negative charges. However, living beings are fundamentally different from inanimate solids.

When addressing the metaphysical world, people often envision a realm completely separate from the universe. This perspective is rooted in thousands of years of thought and has been conveyed to us through religious teachings handed down from our ancestors. We humans continue on this path, mainly by imitating those who came before us.

Moreover, missionaries, scientists, physicists, and philosophers have yet to achieve the scientific understanding necessary to explore metaphysical issues and address human beliefs and convictions thoroughly. While divine texts touch upon the metaphysical world, they do so through references rather than scientific arguments.

Today, with advancements in science, metaphysical claims lacking a scientific basis are not accepted by scientific and philosophical thinkers. Both religious missionaries and scientists struggle to provide scientific arguments against human beliefs and the concept of the metaphysical. Similarly, those who deny the existence of the metaphysical also fail to offer scientific arguments to refute these beliefs.

We understand that the universe came into existence based on the scientific knowledge available to us. If individuals are unable to respond to certain questions or if divine texts contain scientific inaccuracies, it does not indicate a flaw in the creation system. Instead, it highlights our limited understanding of the science behind existence and the insufficient scientific explanations found in these divine texts for humanity.

It was previously discussed that the universe was created within the metaphysical realm, which can be understood as nothingness or the world of quantum vacuum. Now, the quantum world has merged with this concept of nothingness, and although nothingness exists alongside us, we often fail to recognize it. Unlike the material world, which has both a beginning and an end, nothingness is infinite and does not possess these limitations. Within nothingness are entities or "cells" that exist outside the constraints of time.

The difference between the metaphysical world and the physical world lies in the nature of each. The metaphysical world can be described as a quantum vacuum, which means it is devoid of any objects. In contrast, the physical world is characterized by the passage of time and consists of various dualities, such as positive and negative charges, male and female, attraction and repulsion, beauty and ugliness, and light and darkness. These elements are governed by time and ultimately fade away.

In a time-controlled universe, the principle of conservation of energy is rejected. In this case, energy is created from matter, which is transformed into smaller particles of Aliums and photons before being destroyed. Once this transformation occurs, the energy cannot be converted back into matter, leading to the eventual destruction of the universe itself.

In this discussion, philosophy and quantum physics have reached a consensus: the universe cannot emerge from nothing on its own. Therefore, there must be a factor that exists outside of time and is not subject to its control. Additionally, this factor possesses a framework that can transform the concept of zero into both infinite positivity and infinite negativity, leading to the realm of Tanerium and the bipolar world of time control.

When discussing science, we inevitably encounter the concept of intelligence, which is evident throughout the universe. Intelligence exists not only in the world but also within us as humans, allowing us to question the universe and even the Creator. However, according to the principles of physics, this intelligence must have a reference point—the Creator of the universe, who resides within us. Like the universe, we are enveloped in this intelligence.

It is misguided to view ourselves as challenging the Creator; in reality, we are challenging our own nature as intelligent beings derived from the Creator, much like a drop taken from an infinite ocean. This drop symbolizes intelligence, with its components present in every living being, often referred to as the soul based on its structure.

The soul can expand through physical changes and reflects our actions through its "coloration," which can signify good or bad qualities. While our actions do not directly penetrate the soul, they act as a veil that can hinder the flow of graces to those souls that are darker, leading to wrongdoing. In contrast, souls associated with positive qualities lack such barriers, enabling them to receive graces and pursue goodness, kindness, love, and humanity.

The Creator, referred to as the infinitely super intelligent network, continually offers the grace of guidance. Some individuals embrace this grace, while others neglect it. Ultimately, it is a personal choice; individuals determine whether to engage in good deeds or to act poorly, thus forfeiting this grace. It is crucial to understand that this process is governed by law and order; the Creator does not intervene in it.

I previously explained that the soul is associated with intelligence, just as the Creator embodies intelligence. When you are asleep or unconscious, you are not considered to have a soul. However, as soon as you wake up, you become a soul. A body without a soul can be healthy, and its organs may function properly, but it lacks feelings, movement, and will.

What is Reason?

Reason is essentially intelligence and can be regarded as humanity's very soul. It serves as humanity's best guide, being the closest, fastest, most compassionate, and most economical option available. As a blessing and divine gift, it can be seen as the ultimate guidance for humans.

Reason embodies the Creator within humanity's physical form. Trampling on reason is trampling on the divine, reducing oneself to an animal that only seeks to satisfy animalistic desires. A person who turns away from reason deprives themselves of their own will and agency.

It is crucial to remember that each individual is responsible for their own actions. This is a fundamental law set by the Creator of the universe, and it is immutable. No one can alter the fate or destiny of another person; only the individual can do that for themselves. Therefore, do not interfere with the natural order of existence. Each person is ultimately accountable to themselves and can only offer guidance to others in specific situations—forcing beliefs or actions on others is not within your rights .

When you attempt to impose your views on another, you may not realize that their actions and experiences shape the understanding of each individual. If you insist on coercion, I argue that such behavior reflects a lack of understanding and that you may be acting foolishly or irrationally.

To exploit nations, colonialists seize control of their intellect and wisdom by promoting superstitions. They manipulate people like animals, directing them towards their personal interests. No form of colonization or exploitation can surpass the intellectual domination of humans. Those who surrender their intellect to exploiters become intellectually weak and are indoctrinated into believing they lack the power to think and understand.

Exploiters of intellect and thought often reject the concept of God and deny divine justice. Humans should be mindful of the natural laws and the system of existence, which is precise and calculating. In contrast, the intelligent network is inherently compassionate. If you conduct yourself with honesty, avoid meddling in others' affairs, do not harass anyone, express regret for past misdeeds, assist those in need to the best of your ability, focus on the present rather than dwell on the past or worry about the future, and maintain gratitude for what you have received without demanding too much, you will find yourself closer to the intelligent network.

Ultimately, this intelligent network resides within your own being; there is no physical distance separating you from it. Instead, the distance is metaphysical.

The Creator, or the Intelligent Network, shares its knowledge with those who are similar to it. A key characteristic of the Intelligent Network is its kindness and love, which is reflected in its provision of sustenance. However, this network does not interfere with human lives. The Creator, or God, does not see Himself as the guardian of humanity; instead, He has granted people the power to shape their destinies. The rewards and consequences of their actions depend on their choices. This is the order and law established by the Creator, who does not intervene in it.

Humans have two types of desires: physical desires and desires of the soul. For centuries, many have mistakenly believed that material desires stem solely from the body. However, this perspective is flawed. The human body does not possess the power or authority to make decisions. Instead, all desires of the soul, which seek to enjoy life, are governed by the passage of time.

The analyses and decisions made by the soul, or intelligence, are executed through the body. The brain, along with all the body's organs, serves as a tool for the soul or intelligence. Since the soul is metaphysical, it cannot operate or experience pleasure in the material world without the physical body.

All living beings, including humans, rely on intelligence; without it, neither the brain nor the body holds any significance. Information is stored in the intelligence system, and this information can also be recorded in the human brain. However, can this information be accessed without the presence of intelligence?

Second, there are metaphysical desires that stem from the universe's intelligence network, guiding us toward perfection. In contrast, desires for leadership, status, wealth, and traits like prejudice, pride, arrogance, theft, deceit, jealousy, self-righteousness, and revenge are rooted in physical and material concerns. These desires enter our consciousness, urging us to seek material pleasures. Ultimately, the choice of which path to follow is yours.

Moreover, the intelligence of the universe, or God, resides within you. Just as you communicate with others, take a few minutes each day to connect with

God. Focus on the present moment, free from material thoughts, and without dwelling on the past or worrying about the future. Make a covenant with God and remain true to it, as He values your actions over your words.

Avoid elevating material desires because they are tied to a destiny you have created for yourself. Remember, the Creator of the universe possesses a level of intelligence far beyond that of any individual you may converse with. While that person may overlook or misunderstand what you say, the Creator is fully aware of your thoughts, origins, and true essence.

Instead of seeking solace from the Creator or appealing to Him, do not turn to His creations. This is a sign of ignorance and folly. Another birth follows every death; intelligence does not vanish; it is merely the body that ceases to exist. The reason people don't like the concept of death is due to feelings of distance and separation.

Millions lost their lives during the Crusades, a conflict between Islam and Christianity. Additionally, in the Middle Ages, priests in Europe often condemned scientists to death, as illustrated by the case of Galileo. In those times, no one in Europe dared to speak about science, for the church would swiftly condemn them to execution or face a firing squad.

Humans on Earth often live in ignorance. They fail to realize that if they kill another race or nation in a particular region, they will be reborn in that same area after death and will encounter countless misfortunes, misery, and suffering. This concept is rooted in quantum physics and is fundamental to the nature of existence. Humans do not simply perish, rather, they are part of an intelligent structure that constantly changes bodies. Every action they take has consequences, and they will have to confront those consequences in their next reincarnation.

You are responsible for your own actions. You cannot be held accountable for others, nor can you be punished in their place. You are not beheaded, tortured, or transformed into an animal because of someone else's actions. The intelligent network—or God—is silent and does not intervene. It's time to come to your senses! No one has the right to interfere with or change the fate of others, whether individuals or nations. To interfere in this way is to challenge the Creator of the universe. Those who hold this mindset typically reject the natural order established by the Creator. They may, knowingly or unknowingly, consider themselves above God and seek to alter the system according to their own desires, claiming greatness with pride and arrogance. These attitudes reflect a kind of lowly humanity.

God says, "I will not change the fate of a people unless they wish to do so." He has placed the fate of humanity in their own hands and granted them authority, making each person responsible for their actions. Indeed, an individual cannot manipulate the system of existence. Those who attempt to do so will not only fail but will also meet a disgraceful end, for God is no joker.

Unfortunately, in many backward societies, some individuals believe they are responsible for others. Instead of providing help or guidance, they consider themselves representatives of God. They are often so misguided that they fail to realize that even God does not appoint himself as the guardian of humanity, despite being omnipotent and the Creator of the entire universe.

What I shared with you was not conveyed by religious leaders or prophets, as they lacked knowledge of quantum physics and the philosophy of the universe's structure. While I still have much to learn, what I told you is the truth. The universe is governed by order and laws; it is not a place where everyone can do as they please without consequence. God cannot be deceived. You know that all existence, including humanity, originates from God. When you oppress anyone, you are ultimately oppressing yourself and God. Do you humans believe you are separate entities? Quantum physics and science have demonstrated that all existence and nothingness stem from the Creator Himself. There is nothing other than Him.

Unfortunately, ignorance on Earth is leading humanity to destruction. Surprisingly, this ignorance has peaked among heads of state and even some scientists, with philosophers being the notable exception.

Philosophers are compassionate leaders of human societies and saviors of the planet. Among physicists and scientists, some are very inclined towards philosophy and strive to save human society. The leaders of most countries, driven by material desires, position, and status, take their nations hostage to achieve their worldly ambitions, leading the world into conflict and war. They kill millions of people through religious, ethnic, or national propaganda. This method has been prevalent on the planet for years and is being utilized in a newer style today.

Today, advancements in medical science and healthcare have significantly reduced human mortality rates. However, challenges such as population growth, dwindling food resources, the unchecked ambitions of powerful nations, uncontrolled population increases in some developing countries, and a lack of proper education are driving society toward ignorance and moral decline.

Unfortunately, some politicians with animalistic tendencies employ scientists who share similar traits to create weapons of mass destruction, resulting in misery, devastation, and the degradation of human values and dignity.

Some media figures, driven by animalistic instincts, align themselves with ruthless rulers and undermine the very foundations of humanity. Upon reflection, it becomes evident that certain individuals have adopted predatory behaviors. Unlike a predator in the wild, which hunts and consumes its prey solely for sustenance and then moves on, these two-legged creatures—humans—kill thousands of their own kind and are never satisfied. This stark reality raises questions about humanity's nobility, revealing a disturbing potential for humans to sink to the level of the lowest creatures.

The rapid advancement of science, coupled with the control of profit-driven capitalists, has compounded humanity's challenges. The rising influence of science, alongside the prevalence of profit-driven capitalists, has significantly contributed to the challenges humanity faces today. Currently, the world is engulfed in confusion, disorientation, indifference, oppression, and violence against individuals. In the past, people acted with consideration for one another, often driven by fear of the Day of Judgment. However, we now observe that many who profess to be religious leaders and representatives of various sects seem to lack a genuine belief in God and the Day of Judgment, frequently displaying some of the worst moral corruption.

Every day, millions of tons of oil and gas are consumed worldwide, with a significant portion used to produce weapons of mass destruction. This process contributes to air pollution and exacerbates human issues. The lack of logical thinking among people is widespread. If these resources were redirected toward development, we could transform deserts into arable land. This approach could eliminate the need for basic food supplies and help manage the planet's population. Additionally, improving education would alleviate ignorance and benefit many individuals.

Informed individuals, unlike the uninformed, make positive contributions to humanity.

The great powers in the international community often disregard the rights of weaker countries. They justify their bullying actions by invoking the outdated and harsh law of the "veto," a concept that dates back to the Paleolithic era. How can the leaders of nations with veto power claim to uphold humanity both in society and within their own families while trampling on the rights of weaker nations? This raises a profound question: how can we say that humans are the noblest of creatures? In my view, the universe is not as chaotic as some believe; instead, it exhibits a particular order. In this life, a human may find themselves transformed into an animal in the next, stripped of the ability to act or influence their circumstances.

Politicians have enacted a law known as political immunity for themselves, which is one of the most problematic types of legislation. This law allows officials and politicians to act without the fear of legal repercussions, meaning they cannot be summoned before judicial authorities until their term of office ends. In essence, this law grants officials unchecked power to act as they wish, and it should be abolished entirely. Such a law has no place in a just society. Whenever an official commits an illegal act, they should be held accountable and arrested.

Furthermore, there should be a decentralization of power away from governments; political, social, economic, and military authority should not reside solely within the hands of governmental bodies.

The authority of the police or security forces should not fall under the control of the government, as this could lead to the suppression of the people at any time. The leadership of these forces should be chosen by the people through a popular vote. They must carry out their responsibilities in accordance with established laws and regulations. If they commit a crime, they should be promptly removed from power, and a replacement should be elected. Additionally, they should not have political immunity.

Every country should have two governments. The first government should act as an agent, holding the country's political, military, and economic power. The second government, referred to as the deterrent government, should represent the people and be equipped with light weapons. If the first government forces the people or enacts laws that go against their will, then the deterrent government can mobilize ordinary citizens who have been trained in advance to defend the rights of the people against oppression from the first government.

The next issue is that the head of the judiciary and judges should be elected by popular vote. Judges and the head of the judiciary should not be appointed or conspire against the nation's interests while in power. They should not be immune to scrutiny. Additionally, heads of state and officials should not have armed bodyguards; they should behave like ordinary citizens. Everyone should have the opportunity to speak with heads of state and share their ideas. If this procedure is implemented in society, it will restore peace, comfort, and transparency.

People need to impose limits on politicians and government agents. There should be a deterrent government that can protect the rights of the people and prevent the decentralization of power; otherwise, it will be difficult to maintain control. No government should be granted complete authority. Political and judicial immunity should not be extended to anyone, even if they hold the highest office in the country.

Nations, like states, must possess deterrent power. It is time to end the outdated governance methods that have persisted for thousands of years, characterized by oppression and coercion. Nations have the right to possess power and deterrence, and this principle must be upheld. We need a new world order that replaces these archaic governing techniques. The fundamental right of the people is for states to serve them, rather than nations being subservient to the state.

Time Travel

Humans possess individual intelligence or individual souls, which exist within a larger framework of general intelligence or a collective soul. However, why can humans not communicate with the universe's general intelligence? When a material subset forms and takes on a complete physical body, it gains the capacity for individual intelligence. For instance, a human baby or other animals exhibit individual intelligence, having the ability to analyze and respond to their environment. In this context, we can say that they are living beings.

The spirit or intelligence that enters a person grants them authority. In childhood, the human brain is not fully developed and cannot effectively communicate with the universal spirit. At this stage, the newly entered spirit lacks life experiences and, whether willingly or unwillingly, adheres to the demands of its material and physical aspects. During this time, a person accumulates functions for themselves.

As the individual spirit gains experiences, its connection with the universal spirit diminishes. The individual begins to recognize their mistakes and strives to correct their actions, aiming to shed their darkness and return to a state of purity. However, these darks do not easily disappear. A person reincarnates multiple times, gaining enough experience that they eventually tire of misguidance and wrongdoing against others. They learn to show respect, return to the rightful path of humanity, and regain their original self.

By purifying the individual soul and connecting with the universal soul of the cosmos, one can reach a state of heightened consciousness and awareness. In this state, individuals become aware of both the past and the future, gaining insights that are not visual in the conventional sense but are perceived through a deeper level of intelligence. This implies that individuals are linked to the universal intelligence, allowing them to access knowledge beyond the limitations of physical existence. Even without a material body, one can still gain this understanding.

Seeing the past does not suggest that one literally travels back in time. Instead, it reflects the understanding that the universe, known as Tanerium, is governed by a single program. What individuals perceive is the fundamental program of existence itself. This universe has undergone creation and destruction countless times—billions of iterations—and will continue to do so. Therefore, a person who has attained such intelligence can also glimpse the future, as it is part of the same overarching program of existence.

There is a discussion of hellfire in religious texts. Humans have always feared fire, and in ancient times, there was no scientific understanding of how the soul changes bodies or what being accountable truly means. Even if such knowledge existed, it would have been difficult for people to comprehend, leading them to discuss fire as a consequence.

It's important to note that there is no place in God's creation for a crematorium. Furthermore, we humans should not mistreat or oppress animals. If we do, we may one day be forced to return to a more primitive state, facing the repercussions of our actions.

The entirety of existence is a manifestation of God. We are all immersed in God, often without realizing it. If you close your eyes and allow yourself to feel God's presence, you may come to understand that you, too, are part of the divine. There is no separation; you exist within the Creator, and the Creator is full of love and affection for all of existence.

The world is progressing toward perfection

Every living being that breathes is on a journey toward perfection. While the material body is perishable and ultimately temporary, it is the individual intelligence or soul that truly strives for perfection. Earth can be likened to a school, where we advance through various levels—elementary school, high school, and university. A student enters this school and, after nine months of study, advances to a higher class. If, during this time, a student falls behind in their studies or changes classes, they must make up for the lost work during the summer to progress; otherwise, they will remain in their current class.

This process of advancement is also reflected in the stages of the soul's journey as it progresses through successive bodies. Initially, the soul gains experience in the bodies of smaller living beings as it begins its existence within the realm of time. Through many cycles of reincarnation, possibly thousands, the soul ultimately evolves and reaches the human form.

The soul or intelligence that enters the body of a living being initially lacks experience in the realm of time and often seeks material desires and worldly pleasures. What many humans do not understand, and what I am revealing for the first time in this book, is the true nature of the soul. This soul enters the human experience after undergoing countless experiences, allowing it to pursue desires that align with universal intelligence, ultimately leading it toward perfection. This journey also enables the soul to satisfy the material desires of the body.

Essentially, the soul can nurture qualities that foster existence, such as love, compassion, and appreciation for life, ultimately striving to reach their highest forms. This process is inherently structured within the system of existence; no soul is compelled to enter the realm of time. Instead, this journey is entirely voluntary. The soul's greatest desire is to emerge from the absolute darkness and silence of nothingness, seeking to experience the pleasures of the material world.

The souls that enter the human body often lack sufficient experience because the biological systems, brains, and abilities of humans are far superior to those of animals. In this context, making mistakes is a regular part of being human. Therefore, individuals should not be tortured or persecuted for minor errors; they should be forgiven so they can learn and grow. Execution is not a justifiable response.

Those who advocate for the punishment of human souls may be ignorant of the complexities of human societies. Interestingly, they may find themselves facing similar fates or even worse outcomes because a fundamental law exists: everyone is accountable for their actions.

When someone curses or slaps you, it is essential to ask for forgiveness for them rather than responding in kind. Aggression often stems from past actions, and if you retaliate, you are likely to perpetuate a cycle of harm. Instead of achieving personal growth or fostering societal reform, you will only contribute to a cycle of revenge.

Ultimately, for this cycle of retaliation to end, someone must choose to forgive.

Did you know that if you don't violate someone's rights, it becomes impossible for them to violate yours? Unfortunately, many people are unaware of this principle. Imagine you're an 18-year-old young person living in a family where your level of humanity resembles that of a university student, and you have a strong sense of spiritual awareness. In contrast, your family members may only be at a first-grade level in terms of humanity and have a lower level of awareness, often making life difficult for you and others.

Don't be surprised! You might encounter such individuals at school, university, work, or even among your friends. It's not unusual at all. In these situations, it is essential to be tolerant and patient. Strive to be a positive example and teach them about the value of humanity. If that feels overwhelming, try to avoid struggles and quietly distance yourself from their negativity.

Weak and ignorant individuals often react impulsively without thinking. In contrast, great individuals are forgiving and generous, while ignorant people frequently display spite and a desire for revenge. The greatness of the human spirit, as well as our awareness, is shaped by our essence, which is influenced by our past and present actions.

God does not play dice!

The great scientist Albert Einstein said, "God does not play dice with the universe." Einstein suggested that randomness plays no role in the laws of the universe. Nothing happens by chance or coincidence; rather, the world we inhabit is infinitely intelligent. Although this intelligent network operates based on its intelligence, it follows the fundamental laws of existence and science.

Sometimes, you may interact with someone and behave in a particular way, expecting a specific reaction. However, when you compare their response to your prediction, you might find it completely unexpected. Upon reflection, you may realize that this person chose the best response, one that your current understanding hasn't yet reached.

The infinitely intelligent supernetwork selects the optimal path, as our scientific understanding does not extend to its level of insight, leading us to perceive its actions as mere coincidence or accident. In relation to the ICFO[1] experiment, I emphasize that the speed of light, which we measure on Earth as 300,000 kilometers per second, is not the actual speed of light. This measurement reflects the difference between the speed of light from opposing points in the universe. The fundamental reason behind the speed of light is the gravitational influence of the superlayer of Paravazium from beyond our universe.

When light is produced at a point, it spreads out in six directions, forming a sphere of frequency. The universe can be imagined as a spherical shape surrounded by the superlayer of Paravazium force. At the central core of the universe, the speed of light is effectively zero. At different locations within this sphere, the speed of light can vary, influenced by the differences in gravitational forces from opposing points. Furthermore, any particle moving close to the speed of light experiences a reduction in its gravitational force in the direction of its movement. Therefore, we cannot expect an electron, whether positively or negatively charged, to possess the same initial properties. While we know that electrons orbit the nucleus, their rotational speed is significantly lower than the speed of light. In my opinion, the statement "God does not throw dice" is accurate; all events and occurrences are rooted in scientific principles, even if we do not always understand their underlying scientific origins.

Heisenberg's uncertainty principle is unacceptable. Everything in the universe is based on scientific principles, and the scientific method provides a solid foundation for understanding. However, as human beings, we often encounter

[1] The Institute of Photonic Sciences

challenges in attaining that level of certainty, which leads to questions about chance, coincidence, and uncertainty.

Based on my research, I assert that no speed can exceed the speed of light. It is essential to consider that we may be in a region of the universe where the speed of light appears relatively uniform in six directions, making it difficult for us to perceive any differences. However, if we were near the center of the universe, we might find that the speed of light there is effectively zero.

Is there a way to compensate for negative actions?

Yes, there is. Everything is in the hands of the individual. God never leads a person toward evil. There is a continuous divine guiding light in nature that directs humans toward awareness and serves as a reminder. However, those who engage in negative actions often find it difficult to perceive this light due to the dark layers that cloud their minds. These harmful actions obstruct their ability to access guidance.

There are individuals who, when faced with the same guiding graces, awaken to their situations and choose to change their paths. As religious teachings suggest, they embark on a journey of repentance and express regret. In this process, they interact with a network of superintelligence, which helps them distance themselves from the dark influences surrounding their intelligence or soul. This allows the guiding light to reach them continuously, enabling them to embrace a new path toward humanity.

During this time, the intelligent network pays close attention to their actions. While the good deeds they perform help mitigate the impact of their prior wrongdoings, it is essential to understand that these actions are not complete forgiveness. Patience is crucial as they navigate a complex journey filled with challenges, and they must come to terms with the consequences of their past. If a person were to die during this phase, they would have to confront the same stages in their next life.

Many people feel dissatisfied with their lives, but they may not realize that they are paying the price for choices made in the past. They often blame the Creator or God, thinking that God is hostile toward them. This perspective lacks scientific reasoning, and many are unaware of their own history. When something is taken from them, they tend to blame themselves for past mistakes.

Ultimately, everything is within a person's control. God provides guidance, but it is up to each individual to choose their path and shape their destiny.

Today, many people reject the idea of divine justice primarily because they lack knowledge of metaphysics. The secrets of existence have never been fully accessible to everyone. Those who understood this science in the past not only were unaware of quantum physics but also lacked the strength and courage to articulate their insights clearly. This was mainly due to superstitious individuals labeling them as atheists or agnostics.

With this book and its accompanying speech, readers will gain a clear understanding of the correct approach to comprehending God and justice, the nature of God and the soul, and how to distinguish between obligatory, voluntary, and accountable destinies. This understanding can lead them to stop committing crimes and oppressing others, enabling them to pursue higher ideals of humanity and their happiness. Nothing of this nature has been presented scientifically before. Science is undeniable, whether one is a believer or an atheist.

Although I've come to accept it as usual, I'm often surprised by how ignorant humans can be. Many do not realize they are living in an intelligent world—an intricate network with infinitely vast dimensions—while their understanding remains limited.

It's astonishing that despite this ignorance, people are unable to prevent their mortality, even for a moment. Yet, with great pride and arrogance, they issue commands, impose restrictions, and even make decisions regarding life and death.

Individuals who repeatedly break their promises and continue to make the same mistakes send a message to this intelligent network, which responds by manifesting their harmful actions and affecting their lives accordingly.

We live in a world of superintelligence, yet we often fail to recognize it, believing ourselves to be the most intelligent beings. This belief is misleading.

What gives humans reasons to be proud and arrogant? Is it the strength of our bodies, which are ultimately vulnerable to destruction? Is it our wealth and possessions, which we cannot take with us after death? Is it our social status and positions that can be lost at any moment? Or is it our intellect, which is part of a larger network of intelligence? If a person goes to sleep and never wakes up, they lose everything, including their body. The only valid value and currency in life comes from our good deeds. Each individual is solely responsible for their actions, not those of others.

How did the universe come into being from nothingness?

Science tells us that before the universe existed, there was no time, no space, but it was quantum vacuum. I spent considerable time researching and investigating the nature of nothingness. I was fortunate to receive guidance from an intelligent network of existence that paved the way for me on this critical journey. I came to understand that the universe emerged from nothingness, which I recognized as a quantum vacuum; yet it is not absoutely empty.

Furthermore, I realized that spirits originate from nothingness and consist of various components. This leads me to believe that nothingness itself is composed of elements that possess infinitely higher intelligence. Additionally, the first mass that came into existence in the universe manifested as a superatom called the superatom of Tanerium. We know this superatom was initially created within nothingness itself and eventually paired its structure to form the universe.

We arrive at the notion that nothingness is comprised of atoms with seven layers and a central nucleus. However, unlike traditional atoms, these are not paired. This lack of pairing means they are not governed by time, and spirits fall into the same category. I have named these atoms of the realm of nothingness "Paratium."

The key difference between atoms and Paratiums is that Paratiums are not only free from the constraints of time, but they also possess vastly greater metaphysical intelligence and are capable of self-creation. Paratiums are essentially particles of God. These particles constitute the soul and are fully present in both the entire realm of nothingness and the current existence. This is because the universe of existence was created within the very essence of nothingness itself.

The intelligence of the universe has united vast masses of particles, known as Paratiums, to create a super Paratium called Tanerium from nothingness. This process has transformed it into a pair of entities, and through the super layer of Paravazium, it grants the universe the ability to dominate time. We humans are still astonished by the magnitude of this concept. Each of those Paratiums has formed a galaxy within the realm of time domination. Atoms, in this framework, are solid particles subject to the influence of time.

Paratiums have only one decision-maker and one singular source of power. The structure of the soul in animals is composed of Paratiums. Smaller animals have a lesser quantity, while larger and more intelligent animals, such as humans, contain a greater amount of these particles.

The entire universe, encompassing both nothingness and the present, is composed of entities known as Paratiums. These Paratiums are self-created and possess infinite intelligence. They exist entirely in both the universe of nothingness and the current universe, transcending the limitations of time and space. Consequently, the laws that govern the structure of the universe do not apply to them. For the Paratiums, there is no past or future; they have no beginning and no end. Everything that exists, both externally and internally, whether first or last, is eternal and everlasting.

Why did God create the universe and humans?

Let me illustrate with an example. If you were to describe two environments, the first would be absolute silence and complete darkness, where there is no movement, speech, or reaction. The second environment, in contrast, is vibrant with movement, filled with the pleasures of life, socializing, and interactions among people. It's likely you would choose the second environment. In the first environment, your greatest desire would be to escape into the second.

The first environment represents nothingness, while the second exists within the realm of time. All conscious beings long to enter this domain of time and leave behind the void of nothingness. The souls of all living beings, including humans, have willingly chosen to enter this world to experience these joys.

In this life, you cannot blame God for your troubles, as you have the agency to make your own choices. He has bestowed upon you reason and freedom, allowing you to shape your destiny through your actions. His role is to guide you, not to dictate your path.

Most people view life in the universe as a source of pain and suffering, but it is not. Life is more like a university that provides us with knowledge. Education and knowledge often come with their share of pain and suffering. When we refer to humans, we speak of the soul, not the body, which is constantly changing and serves merely as a vessel or tool. The soul exists in nothingness—in absolute darkness and silence. Although it possesses intelligence, it is also entirely ignorant, lacking awareness and experience of material life. Souls can attain consciousness only in the material world and through the challenges of life. Every problem that arises brings with it an opportunity for awareness, and humans should regard these challenges as treasures to be used in later stages of life.

The intelligent network of the universe, or the Creator, does not grant privileges without effort. One must work hard, and there is no discrimination in this regard. The universal network seeks awareness, which is enhanced through human experiences. When even one individual attains awareness, that knowledge is transmitted throughout the entire system of existence. This is why the Creator continually helps and guides humans toward higher awareness. Conversely, when individuals or groups turn away from knowledge, that ignorance is also transmitted throughout the cosmos, causing displeasure to the Creator and potentially leading to disaster.

You humans are from God, and in a sense, you are creators yourselves. Consider this: when one person or a group commits acts of violence against others, you are compelled to either stop them through force or confine them.

The Path to Awareness

I have mentioned this point repeatedly, and I will reiterate it: the Creator has brought all of existence into a state of time control, deriving from His existence and based on the principles of quantum physics. The entirety of existence is a manifestation of the Creator.

Humans are intelligent beings, and this intelligence reflects a form of superintelligence that stems from nothingness, or the Creator Himself, which operates in a state of time control without change. This intelligence is like a drop in an infinite ocean. When this intelligence enters time control and incarnates in a physical body, it may change bodies many times; however, it still lacks sufficient experience and tends to seek material pleasures.

Right now, I, as an intelligent being, am speaking to you, who are also intelligent beings. You possess the power to direct that intelligence toward a path free from wrong, unjust, and unreasonable choices. You can walk the correct path within the network of the universe's superintelligence, striving for freedom and liberation from the absurd dependencies of life. Ultimately, you wish to follow the rightful path without deceiving yourselves.

Humans possess intelligence that can lead to both good and evil. Pursuing material pleasures without sufficient experience in managing time can lead you down a path of evil, causing harm to others and resulting in negative consequences for yourself. Recognizing this, you should choose a path that respects the rights of others to avoid suffering. The only accurate way back to your origin is through the Creator.

Your authenticity reflects that of God, a truth often overlooked. The Iranian mystic Mansur Hallaj declared, "I am God," but his profound statement was misunderstood, leading to his execution under the Abbasid Caliph. Similarly, Jesus Christ claimed, "I am the Son of God." Even had he said, "I am God," few would have believed it, jeopardizing his mission and ultimately leading to his demise.

What is a dream?

The question of what constitutes a dream has fascinated humans since the dawn of time, yet no one has been able to provide a scientific answer. For thousands of years, people have referred to a concept called the "mind" in various languages, attributing all aspects of life to it. However, the reality is that there is no separate entity called the mind; instead, all human brain cells function through an intelligent system, which has been described previously in relation to the concept of the soul. Without this intelligence or soul, the human body is merely a physical entity. When we are asleep or unconscious, this intelligence or soul temporarily detaches from us. However, it continues to oversee the body, ensuring it remains alive and can eventually awaken and move again.

The body is a tool at the disposal of intelligence or the soul, which, through brain cells, enables human beings to live and move. As long as the body and brain cells are in good health, the soul or intelligence utilizes them as a compound or vehicle. The soul alone lacks the power to perform physical actions and must rely on the body of an intelligent being. It can only act through the brain cells of this being, using them to fulfill its desires, set the body in motion, and achieve its goals. When the body's health deteriorates, the soul separates from it and seeks another body to begin a new life, aiming to improve and enjoy dominion over time. Intelligence is a form of superintelligence of the universe and the creator that brought all existence into being from nothing; it is not bound by time or space and is eternal and indestructible.

The soul is composed of tiny components called Paratiums, which together form the essence of intelligence and the soul itself. The human understanding of the metaphysical world and the concept of the soul is quite limited; therefore, I will briefly explain these points.

Souls consist of intelligent Paratiums that embody the nature of nothingness as well as the essence of the creator or origin of existence. In their pure state, souls exist in absolute silence and darkness, yearning for the moments when they can emerge into the bipolar world and experience the pleasures offered by the passage of time.

Paratiums lack the experience of living within the domain of time. When they inhabit a body, their primary desire is to enjoy life's pleasures, a desire shared by all humans and living beings. However, the overarching goal of the universe's intelligence network is evolution—specifically, the aim is for souls to gain experience and grow while navigating the realm of time.

Paratiums who have just entered this world often lack experience in time management and tend to seek immediate pleasures. This inclination can lead humans to make mistakes and infringe on the rights of others. While an individual's past experiences play a significant role in their behavior, seasoned and enlightened Paratiums can avoid such errors.

Some believe that the human body inherently enjoys life, but this perception is misleading. The body itself cannot experience enjoyment; rather, this feeling is attributed to the soul. All emotions are connected to the soul, which embodies intelligence. The body serves solely as a vessel for the soul, facilitating the promotion and enjoyment of life.

Souls on Earth not only acquire more experiences but also grow in essence and depth. This expansion occurs through the process of inhabiting multiple bodies over time. Reflect on your own life: have your experiences from childhood been the same as those you have today? Your answer should be "no."

Humans are continuously progressing toward greater understanding and intelligence. We need to draw upon the knowledge and wisdom of the more experienced Paratiums within our souls. We should aim to enjoy life while respecting the rights of others, continuing our journeys with love and compassion for all of existence.

Paratiums are constantly exploring. Some people believe that humans only dream when they are asleep, but this is not the case. Humans also dream while they are awake, although they may not be aware of it. When you walk or sit and rest, shapes and images pass through your mind like a film without your conscious interference; these are all dreams. Inexperienced Paratiums exist within the souls of all humans and can lead to despair, depression, arrogance, and the oppression of others. The best way to utilize experienced Paratiums is by following the path of wisdom and logic. In short, treat others as you would like to be treated yourself.

Humans dream while they sleep, and animals do as well. Dreaming is a playful exploration of the soul and intelligence. There are various types of dreams, and I will focus on two of them:

Childish Play Dreams: These dreams involve the soul creating scenes from the past or future that it wishes to enjoy, resembling a form of playful escapism. Such dreams are pretty standard, and you might encounter similar scenes while awake—a phenomenon known as "sleep in wakefulness." Dreams emerge when the mind is at rest; what comes to mind during active thinking is not a dream. Instead, dreams represent the soul's explorations, free from the constraints of will.

Warning Dreams: In these dreams, the soul often connects with a broader intelligence network of existence. This connection allows the soul to access insights about its future and destiny, revealing essential messages. Without the soul, a person is merely a physical entity. Warning dreams carry specific meanings that depend on their interpretation.

All human functions are linked to the soul rather than the body; the body serves merely as a tool. Talented and knowledgeable individuals can activate a greater percentage of their brain cells, suggesting they possess a strong soul and have significantly transformed their bodies. Consequently, these individuals make very few mistakes.

The only way to achieve awareness is to understand the concept of "I am the Truth" (Ana'l-Ḥaqq). It is not enough to merely speak the word; instead, one must cultivate divine qualities, love, and compassion for all of existence within oneself.

The origin of all existence can be traced back to a superatom. In previous discussions comparing the world of nothingness to the material world, it was explained that the mass of a superatom, which I call "Tanerium," was first formed from nothingness in the form of a monopolar superatom that was not influenced by time. I named this superatom "SuperParatium." Almost immediately, it paired with another through a super-intelligent network and expanded via the superlayer of Paravazium, leading to the creation of the universe.

In the early stages of the universe, the speed at which the Paravazium superlayer retreated—an invisible energy responsible for the universe's expansion—was extremely high, reaching levels of billions of kilometers per second, equivalent to the speed of light. Over time, this speed gradually decreased until it reached its current value. During the first few thousand years, galaxies and stars formed, but it took millions of years for planets to emerge from the hydrogen explosions of stars before entering the realm of life.

It is significant to note that nothingness consists of tiny components, similar to atoms, which I have named "Paratiums." The key distinction between nothingness and the material world is that Paratiums possess a high degree of metaphysical intelligence, are unipolar, and exist beyond the control of time.

The atoms of the material world are bipolar, subject to time, and ultimately non-productive. Over time, they break down into smaller particles and eventually disappear. Existence emerged from nothingness itself, and intelligent entities exist throughout the universe, though we may not perceive them. A quantum vacuum represents the space of nothingness. When the universe of existence ceases to exist, this quantum vacuum remains. Additionally, this quantum vacuum exists within the universe itself. Simultaneously, quantum particles and objects exist within the realm of existence, forming the material world.

How did life arise on Earth?

All objects and particles in nature possess intelligence because they are bipolar. This has already been demonstrated. Atoms, electrons, and even photons exhibit this intelligence. Atoms, driven by their intelligence, tend to combine and form molecules. These molecules, in turn, share the same tendency to bond and create cells, leading to the formation of single-celled organisms. While scientists refer to these interactions as chemical properties, philosophy interprets them differently: all these chemical interactions stem from a fundamental feeling of need.

This is just the beginning! All living things possess a sense of need and intelligence. A substance that lacks intelligence does not experience needs, and it is this sense of need that drives chemical interactions. For reproduction, plants require genes. The genes of both animal and plant organisms function like integrated circuits (ICs) or USB drives, storing all the characteristics of a living being. These genes are intelligent entities that can replicate themselves under the right conditions, returning the original being to the realm of life. Although I am not a biologist or physiologist and cannot explain this in detail, I believe that, based on the principles of quantum physics and the twenty laws governing the structure of existence, humans and all animals consist of both physical and metaphysical dimensions of intelligence.

We have already discussed that the soul possesses intelligence while the physical body exhibits a form of physical intelligence. However, this physical intelligence lacks the capacity for design and analysis, skills that only metaphysical intelligence, or the soul, can fulfill. As physical intelligence performs its functions, metaphysical intelligence intervenes to complete the work of gene design. This process occurs based on essential elements, the specific location, and climate, and is entirely grounded in the principles of biology.

In earlier sections, we addressed the emergence of water on Earth, noting that while scientists search for water in space, life itself originated on Earth and did not come from elsewhere. The primary factor in designing and producing the genes of plants and animals is the intelligent network that governs all of existence. A crucial question arises: How did such a diverse array of species of animals and plants come into being?

Based on my research, I concluded that animals genetically very similar to each other were the ancestors of subsequent animals with both similar and differing genes. This process continued, leading to the emergence of different animal species. Initially, single-celled organisms appeared in stagnant waters, marking the entry of life into the world through genetic changes facilitated by a complex network of small animals. These small animals served as the ancestors of other species genetically close to them. Although these new animals differed from their ancestors, they eventually mated with their kind and reproduced.

The human race is not the first to emerge on Earth; in fact, we have come and gone thousands of times throughout history. I often think that each time a human being enters life on this planet, they may live for about fifty thousand years before disappearing. Approximately a thousand years later, another human appears. Currently, our human race is nearing the end of this fifty-thousand-year cycle. As we progress, humans on Earth are moving toward truthfulness and spirituality at an accelerated pace, influenced by concepts of reincarnation and body replacement. Soon, peace will prevail, and the more savage aspects of humanity will fade away, leading to an end to wars and conflicts.

We are poised for significant advancements in metaphysical sciences, which will enable us to travel to other planets in seconds. This knowledge already exists in the universe, even if we are not yet aware of it. Humans are capable of conquering the universe, provided we embrace metaphysical science and elevate our understanding. In the metaphysical world, individuals constantly travel to and from Earth, though we are unable to perceive them.

Darwin's theory cannot be accepted for several reasons. First, according to the "law of similarity," which is one of the 20 laws of the structure of existence, animals exhibit similarities in genetics and physical structure. Based on the methods of creation, it is unlikely that they would be entirely dissimilar. How could it be possible for them to lack any similarities at all?

The second reason is that we lack intuitive reasoning to support evolution. If evolution were a reality, we would have observed at least one or more instances of it occurring among humans over thousands of years, or it would have been documented in historical accounts. However, there have been no such observations.

An intelligent network has guided the emergence of animals and humans over thousands or millions of years. The similarities among living beings do not necessarily validate the concept of evolution. For instance, we cannot assert that a cat was once a lion or vice versa, as no evidence has ever been found to show that a lion transforms into a cat or the other way around over time. Similarly, a sheep has never been observed turning into a goat. In fact, there has not been a single documented case of one animal evolving into another, and genetic science does not support this idea either, despite the fact that rare exceptions can occur in nature.

While changes and transformations can happen in inanimate objects, this is not the case for living beings with genes. Such transformations are only possible if the new being possesses an entirely independent and healthy genetic structure. This scenario can occur only once, and it is believed to happen through the will of a Creator to reproduce a new living being.

Is it not possible for the intelligent network that created the universe, brimming with grandeur, complexity, and meticulous order, to give rise to an animal with a different genetic makeup, such as a cat, from the womb of a lion, or vice versa? Some people who lack sufficient knowledge identify as atheists and may not have a well-founded reason for their beliefs. They often raise the topic of evolution, claiming that the universe came into existence by chance. This perspective is misguided. The universe and its creator possess infinite intelligence, and everything within it is governed by a meticulously intelligent system, down to the smallest detail. Throughout history, countless humans have been born and perished on Earth, and this cycle will continue. Other living beings have also emerged in this world for various reasons. The universe was created for humanity to strive for perfection, meaning a person who never makes mistakes.

In this book, I present physics in a way that has never been done before by physicists. I delve into metaphysics, the secrets of creation, and the reasons for existence in a manner unlike any prophet has attempted. I am neither a physicist nor a prophet. While some may not believe me, I swear to God that I have tried to demonstrate that all of existence lies within ourselves. Throughout my life, I pursued this understanding, ultimately realizing that I am merely a tool in the hands of the Creator. In this era characterized by the decline of humanity, a waning of spirituality, and the prevalence of ignorance, I believe I have been entrusted with the mission to guide humanity toward greater awareness.

I pledge to the One Creator that as I write these words with tears in my eyes, I am not myself and feel God's presence. His companionship brings me joy. I find fulfillment when I am detached from material concerns and live in gratitude. I ask little from God, knowing that I can take nothing with me at death except my deeds. I cherish all of existence, every person of every race, and all animals. I am grateful for my life here, indifferent to whether I remain or depart, satisfied with God's will.

Remember the concept of "I am the Truth" (Ana'l-Ḥaqq). All that exists and does not exist is God.

MIX
Papier aus verantwortungsvollen Quellen
Paper from responsible sources
FSC® C105338

Printed by Books on Demand GmbH, Norderstedt / Germany